Autism
Answers & Action

Medical Malpractice and Childhood Neurodevelopmental Disorders

Dinesh Danny

Printed in the United States of America

Disclaimer: While the advice presented in this book is true to the knowledge of the author it represents general guidelines. Individual situations might be different and no advice should be construed as being given with regard to diagnosis or treatment of any specific individual condition. Please consult with medical and other professionals regarding your specific circumstances. No warranty is provided implicit or otherwise regarding the use of knowledge or material provided in this book.

First Printing, 2021
ISBN - 978-1-7366047-0-0

More about the author at -
www.dineshdanny.com

Contents

Part II: Autism and TBI at Birth

Part III: The Impact of Vaccination

Part IV: The Learning

Part V The Legal Recourse

Preface

It was a quirk of fate – an accident that happened a decade ago that ended up changing the lives of two brand new parents. My wife and I were blessed with two adorable boys arriving two years apart. But when our first son turned three, we began to notice that while he was reading better than his peers he was not talking as much. He had difficulty responding to simple yes/no questions and was always hyperactive. Tests ensued trying to find out what was wrong, eventually leading to the dreaded diagnosis of autism. Fingers crossed, we hope that our second son would not suffer the same fate, but things took a turn for the worse for him at age three as well, and he too was soon diagnosed as on the autism spectrum.

Like numerous other parents whose children are diagnosed with autism, our lives were turned upside down. This was a complex problem made even more challenging with two boys on the Autism spectrum. Over the past decade I left no stone unturned trying to find the cause of the problem so that we could pursue a solution. Doctors offered a whole host of reasons, mostly pointing at genetic and environmental causes, but those explanations left me unsatisfied. It wasn't until a brain MRI scan of my second son gave me a light-bulb moment and I suddenly realized exactly what had happened. Piecing together events and incidents from the past, I was able to determine that Pitocin abuse in labor as the cause of their condition. Fingers have been pointed at Pitocin before, but here I was, staring at actual proof of the damage caused by the abuse of the drug and dealing with its consequences on a daily basis.

This book is an attempt to bring clarity to the subject from the perspective of an observant parent. The observation horizon needed to arrive at a proper conclusion about the causes of autism in case of a specific child is in years, and impacted parents are clearly the only people qualified to have an opinion on this topic, since they are the folks that have spent the most time observing their kids and know exactly what happened to them from pregnancy to childbirth to diagnosis and beyond.

As a result of what happened, in addition to writing this book I have joined forces with several other affected parents and launched the Autism Parents Initiative (autismpi.org), an organization designed to spearhead a data collection effort for legal and research purposes based on the ideas presented in this book. Our objective is to settle the autism question once and for all and to enable justice for parents and kids impacted by medical malpractice. Our goal with the data collection is to ensure that kids and parents already tormented by their situation don't suffer further anguish by the burden of legal cases, taking away valuable time they can devote to the pursuit of treatments and remedies.

Readers will notice that I do not disclose any names throughout my book, including those of my boys, the doctor involved, etc. That's because my aim is not to tell our story but the all too familiar story of many other impacted families. The names of hundreds of others could easily be substituted and would fit the fact pattern just as well. While this book explains the background and details of my story as it unfolded in the US, everything here except the parts about legal recourse and data collection translate to other parts of the world as well.

Acknowledgments

I would like to thank my wife and my two special needs boys for the time spent away from them while I wrote this book. They are the reason I wrote this book, but they also paid a hefty price in terms of my time and attention on them.

This book would not have been possible without the help of other impacted parents who gladly shared their stories with me, including answering some detailed questioning, helping to establish many of the paths and possibilities arrived at in this book.

I would also like to thank the many wonderful nonprofits and websites out there for the plethora of good advice and knowledge they share, and in particular the many autism and mental health organizations for all the good work they have engaged in, trying to rid the world of autism.

Introduction

One of the most pernicious aspects of autism is that it doesn't just impact kids – the diagnosis strikes unsuspecting parents as well when they begin to notice their kids are not developing like all the other kids around. It knocks down the natural pride and joy parents have about their children and naturally raises the question of why autism struck their family.

Defining Autism

The first step to solving a problem is to accurately define the problem itself. Unfortunately, in the case of autism, this definition has been a challenge of epic proportions and a topic that has been constantly evolving.

The word autism is derived from the Greek word "autos," meaning "self" (like "automatic" means "by itself"). It was first used by Dr. Eugen Blueler in 1908 to describe the symptom of a schizophrenic patient that seemed drawn inward into himself and trapped in his own world.

The word has since been extended to childhood development symptoms where children refuse to interact and learn from the world around them, but instead get "drawn into" themselves. The International Classification of Disease (ICD) coding system developed by the World Health Organization classifies autism under the umbrella of Mental, Behavioral and Neurodevelopmental disorders. The Diagnostic and Statistical Manual of Mental Disorders (DSM) in the US defines autism as a "neurodevelopmental disorder that is characterized by impairments in communication and social interaction as well as patterns of restrictive, repetitive interests and behaviors during early childhood" (DSM-IV). The subsequent edition (the 2013 DSM-V) places more emphasis on the social communication and repetitive behavior aspects.

Yet the symptoms of autism are multiple and varied – in fact so many and so disparate the disease itself is called a spectrum disorder, and is characterized by a set of nonspecific disorders in development. The most significant symptom is the lack of verbal development, which lessens the ability of the child to engage in social interaction. This results in them withdrawing into their own world and can lead to the development of obsessive and repetitive behaviors. Autism therefore correctly understood is a set of developmental disorders beginning in early childhood.

The fact that autism begins right from birth naturally skews the general perception of the causes of autism as genetic. Yet while there are some genetic causes that create the symptoms that fall under the vast umbrella of behaviors covered by the autism spectrum, they only constitute 10% or less of the total cases of "autism-like" symptoms out there. The other 90% of the cases are relegated to "unknown" causes that are currently under intense research – research designed to show that they are genetically caused as well.[1.1] Doctors seem to be trained to reinforce the claim that "nobody knows what causes autism" when what they really are saying is "nobody knows the genetic causes of autism."

The Real Cause of Autism

Unfortunately, researchers looking into autism have knowingly or inadvertently created an elaborate web of diversions that obscure the truth regarding the causes of autism. The presumption that all autism has genetic roots results in the current research model focusing on identifying genes and pathways and mechanisms so complex that even the current scientific community struggles to comprehend them, let alone the average family impacted by autism's effects. While many genetic diseases create brain function deficits, the converse is not true. Most kids diagnosed with autism do not have any genetic disorders.

For a long time, the story of childhood autism in the western world has been a mystery mired in the complexity of brain development. Our understanding of autism is restricted by our limited knowledge of the intricacies of brain functioning and human intelligence. But as much as some people would like you to believe autism is some freak occurrence of nature, the cause of autism is in fact a story of deception, incompetence and moral turpitude of certain parties involved in perpetuating a lie that has managed to confuse the general population for decades. As I will elaborate on over the next few chapters, the overwhelming majority of cases of autism are the result of medical malpractice and drug injury during birth.

A multitude of scientific studies and immense resources have been dedicated to research on this topic, but the truth is well known to certain groups. As I went about talking to other impacted parents and medical professionals trying to uncover evidence for my own situation, I found their experience corroborated my own findings. I also discovered that there is a surprising number of people even in the medical profession who know or have a hunch about the causes of autism but just don't say it loud enough – either because they are not impacted or because there is not enough evidence and it is against their professional pursuit to stand up and say the truth "as is" and incur the wrath of medical boards and other authorities.

While the evidence of Pitocin abuse remains hidden in plain sight, there is a lot of misinformation that prevents the truth from emerging. In this book I explain why the diversions are misleading. I also explain the "double whammy" that affects kids delivered by induced labor which worsens their condition and the exact reasoning behind it.

In this book I have also outlined the effects vaccines have had and their role in compounding the autism epidemic brought on by pre-existing brain injury. I am not in favor of or against vaccines as taking sides only results in inflaming an already charged debate. Instead, I have tried to explain the conditions under which vaccines might be considered risky and capable of stopping brain damage recovery permanently. These conditions are already outlined in every vaccine label, but parents don't get a chance to look into these and pediatricians are only told of the 'mild' side effects. It is something people on both sides of the vaccine argument need to look into.

I know there is going to be an army of naysayers (including physicians and other 'experts') waiting in the wings to attack this book on some irrelevant technicality, like my being only an engineer and not a physician or an academic. Most of these naysayers have conflicts of interest and reflexively defend their fellow practitioners. To them and to the reader I point out that all the material in this book is based on facts gleaned from scientific papers or other publicly available material from sources like the US National Institute of Health. In some cases I have offered plausible explanations for actual observed facts in the real world. I am not looking for academic confirmation of many of these points laid out in this book, since they are all established facts based on overwhelming evidence. I have in some cases deliberately avoided getting bogged down by the complexity of the science involved. The actual answers are truly quite simple. These low-stakes science arguments almost always result in inaction and continuation of status quo which has resulted in high stakes autism diagnoses and associated challenges reaching historically unprecedented highs. Our effort on the parent-side of the equation must be designed to establish and catalog causes of autism and establish legal evidence where necessary.

Proving the cause of autism in a US court has remained an elusive target for many parents, who know in their hearts and minds the malpractice that caused their kids' condition but are unable to come up with the evidence to prove it. Legal recourse for parents and their children who were impacted by Pitocin abuse is both fair and necessary compensation for those lives impacted by medical malpractice, but parents should know that the statute of limitations for proving cases of childhood autism in US federal courts is 18 years from birth. I urge every reader of this book to join forces and contribute in any way possible toward making this effort a success.

Part I

Understanding the Brain

The brain is the command and control center for the whole body, making it the most complex and least understood of all organs. Our brains make us who we are: our memories, our unique capabilities, our personalities, our intelligence, our consciousness and a lot more. The processes that drive and control these aspects of brain functioning lie at the junction of many different branches of science, including biochemistry, biophysics, and molecular biology.

The most significant naturally occurring risk to brain health in a person's life is the birth process. It presents a dual possibility of brain damage from lack of oxygen and from traumatic injury occurring from being pushed out head first. While natural brain processes are only now being understood, the processes of damage and recovery are still under study. Autism is a brain development disorder and it is critical we start our journey to understand its causes with studying the brain.

CHAPTER 1

Brain Development

Autism is an affliction of the brain, so this opens with a pseudo technical exposition on the human brain and the aspects important to brain functioning and development. If you are a parent of a kid with autism you are almost required to become an amateur brain researcher if you would like to understand what your child is going through. Unfortunately, a lot of impacted parents continue to remain unaware of the true reasons for their children's condition in the midst of all the misinformation and diversions out there. A good understanding of the human brain is essential for understanding the afflictions that can come about as a result of disturbance or damage to any of these very delicate and complex mechanisms.

Our knowledge of the human brain continues to improve with time and some of these facts continue to evolve. What's presented in the following chapters is mostly at the structural and functional levels. Even if you know quite a bit of detail about the brain, I hope there are some details here that may be new and help with an understanding of mental health problems and symptoms that have now become a growing concern in the US and other countries in the world. This basic knowledge is also critical to understanding the terminology used by doctors, medical reports and medications when trying to understand the diagnosis and treatment options in mental health challenges like autism.

Childhood Brain Development

Out of all the living creatures out there with well-distinguished brain structures, there exists one critical factor that separates human brains from all others. And that is the fact that the human brain is an unfinished organ at birth. A newborn calf or fawn can walk right within a few minutes of birth, while the human baby is completely dependent on its parents for years. In fact, 18 years of parental care for human beings is the longest in the animal kingdom. That time also correlates directly to the time it takes the human brain to reach full maturity.

The fact that the human brain is not at an advanced stage of development at birth should not be seen as a deficit. That fuzzy undeveloped brain in the early stages of human life is the reason that allows the human brain to learn a lot more from environmental inputs. This prolonged development time also affords humans the higher skills of language, art, higher cognition, memory, personality, and fine motor skills that are ultimately more advanced than the rest of the living beings on this planet. But the processes that lead to all these remarkable skills also makes the human brain the most complex system and the least understood organ in our body.

At the most basic level, while every organ is vital to the human body, the brain is simply the most critical organ. People die when the brain is dead. Being the command and control center, it powers our consciousness, every subconscious occurrence and every conscious movement or thought. This amazing feat is accomplished by about a 100 billion neurons in the brain, working together to make our conscious movements and actions as well as our subconscious and autonomous activities like breathing and digestion possible. Our knowledge of the brain in relation to those higher skills has been constantly evolving and is now vastly improved from where we were a few decades ago.

From the time of birth, the brain increases by about 4 times in volume until age 6. Even then it's only 90% complete, with full structural maturation occurring well past adolescence. In the early childhood stage, there is a high level of experience-based development of the brain. Audio visual coordination, attention span, motor development and fine motor coordination are all learned skills. The acquisition of these skills in turn helps develop the neural network of the brain further, strengthening certain connections while eliminating others. All this growth over a prolonged period of time results in the development of what is arguably the most complex biological unit known to nature.

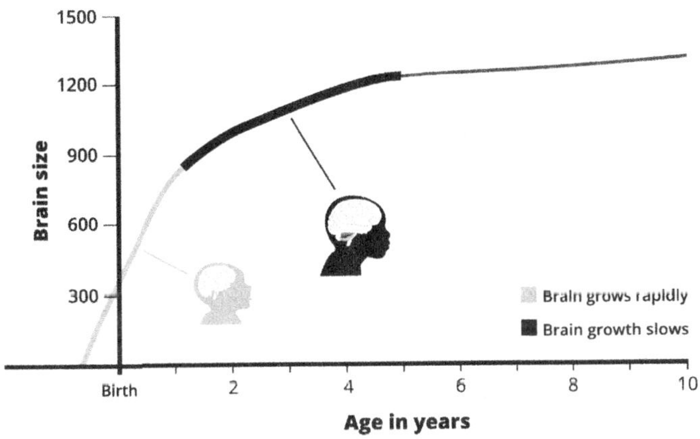

Fig. 1.1: Brain Growth with Age

Stages of Brain Development

Brain development happens in stages, with later stages building on the foundation built in the previous stages. In the first stages the sensory pathways of vision and hearing are established, followed by language and then higher-order cognitive skills such as critical thinking and reasoning. Strong foundations in the early stages makes for better higher-order skills later. The illustrated chart in Fig. 2.2 from the Harvard University Center on the Developing Child details the various stages of brain functional development as a function of the rate of neuronal connections formed with time.

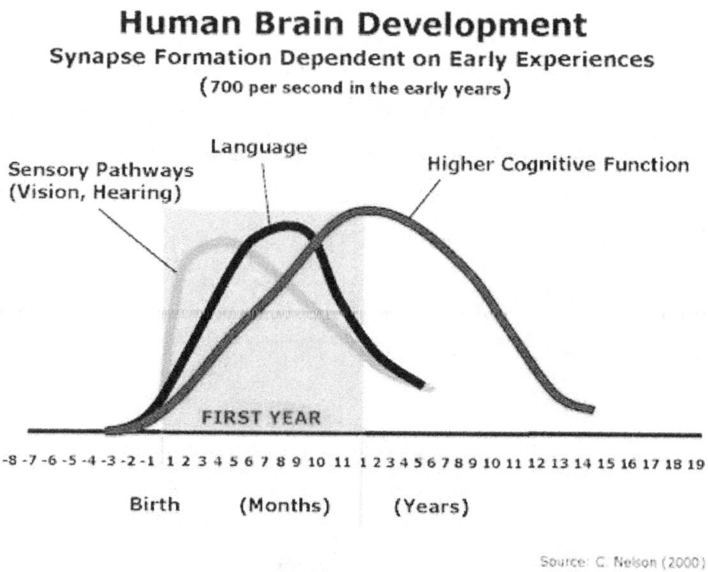

Fig. 1.2: Brain functional Development with time showing neural connections for different brain functions develop sequentially.

Notice how neural connection development really takes off at birth and the maximum rate of development is achieved in the first year of life in a specific order. Experiential learning after birth increases exponentially reaching a rate of about 700 synaptic connections established per second. Neuronal connections relating to the sensory pathways of hearing and vision are the first ones to be established. Language skills are built on top of the sensory pathways of hearing and vision and higher cognitive skills like focus and problem-solving follow language learning. A disruption in the initial development of sensory pathways interferes with the subsequent development of all other skills including language and problem-solving.

Brain Functional Areas

The brain has three main parts: the cerebrum, cerebellum and brainstem. The cerebrum and the cerebellum are divided into 2 lobes each and they are all attached at the brain stem.

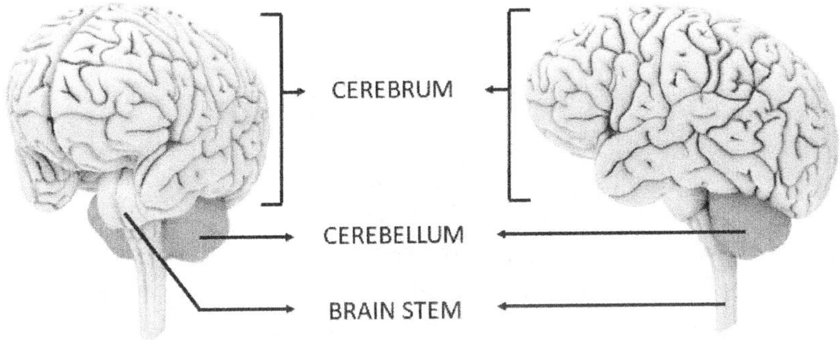

Fig. 1.3: Three main structural parts of the human brain.

The brain stem is the relay center for the Cerebellum and cerebrum taking their messages to the rest of the body through the spinal cord. Additionally, the brain stem is responsible for controlling the autonomous functions of the body including breathing, heart rate, digestion, sleep and wake cycles and many other functions key to the basic functioning of the human body.

The lobes of the Cerebellum are located on the rear side of the head, tucked just under the cerebral lobes. The main known function of the cerebellum is the coordination of various muscle groups to help maintain posture and balance. Shifts in balance are detected and coordinated by the cerebellum to ensure smooth movements. The cerebellum also coordinates eye movements and is involved in motor learning like in learning to ride a bicycle or playing an instrument. Recent research has also suggested a cerebellar role in thinking and mood, although this functionality hasn't been fully explored by scientists.

The cerebrum is the largest part of the brain and is composed of right and left lobes called hemispheres. It performs higher functions like interpreting touch, vision and hearing, as well as speech, reasoning, emotions, learning, and fine control of movement. These are also the functions that are impacted in autism. The following picture gives a good idea of the various areas of the cerebrum and the functionalities they control.

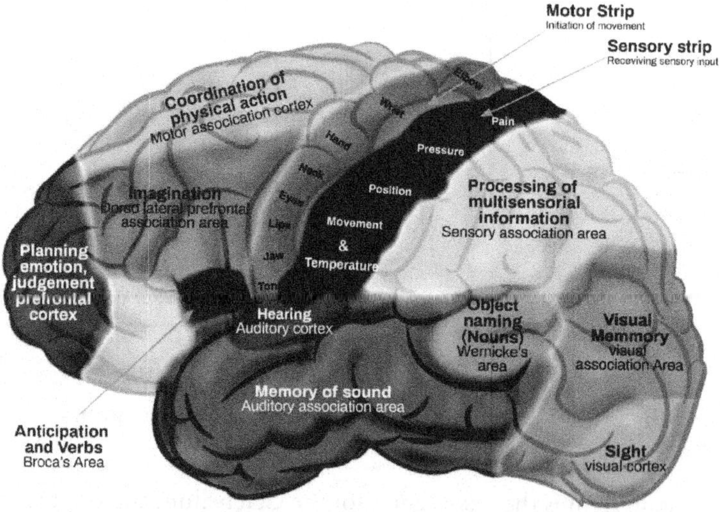

Fig 1.4: Cerebral functional areas

Notable special areas and features of the brain and their functions are shown in Fig. 1.4. Notice how the following main functional areas are surrounded by other areas that support the association and coordination of these senses:

Main Function Area	Surrounding Support Area / Memory
Sensory strip	- Sensory association area
Motor strip	- Motor association coordination area
Hearing	- Auditory association area
Sight	- Visual memory and association area
Planning/ Emotion	- Imagination, prefrontal association

As we will discuss in upcoming chapters, the cerebrum, being the largest, outermost region of the brain, makes it the most susceptible to traumatic damage particularly during birth, when a baby makes its way out of the mother head first. The higher-order functions controlled by the cerebrum are also the same found deficient in kids impacted by the autism spectrum disorder. Deficiencies in auditory and visual processing, lack of fine motor skills and slow processing speeds, are all telltale signs of damage to the cerebrum.

Key Take Away: The human brain is the most complex and least understood of human organs. Brain development occurs in stages, with every stage of development being the foundation of the next stage. The first two years of life represent the period of maximum brain development, and 90% of the brain is developed by age 6. Damage to the main underlying networks of the brain is largely irreparable. Any damage to the brain at birth means subsequent development is significantly hindered.

CHAPTER 2

Brain Functioning

Brain substructures like neurons, grey and white matter, the myelin sheath, and neurotransmitters play an important role in understanding neurodevelopmental disorders. Let's look at these in a little more detail.

Neurons

The neurons that make up the brain are of various shapes, sizes and types that serve varying functions in the brain. Outside of the central nervous system, neurons are found all over the body in the peripheral nervous system, the enteric nervous system and every other part of the body that the brain controls either autonomously or voluntarily. In the brain, neurons work together forming neural networks that are responsible for all the higher-order functions that the human brain is exclusively capable of like speech, learning, thinking and imagination.

Neurons are specialized cells with very distinct shapes that are designed for the creation of networks to pass chemical and electrical signals. The diversity in the structure of neurons is so large these cell types are still being discovered. Unlike other cells, neurons never divide, and neither do they die off to be replaced by new ones. They usually cannot be replaced after being lost, although there are a few exceptions. There are also neuron support cells, called the neuroglia or glial cells, that help regulate the chemical environment of the brain. The main signaling neurons are thought to be of the most common multipolar neuron structural type, with a branched cell body and a single elongated cable-like structure called an axon, which also has a branched end to it. The human brain is estimated to have about 100 billion neurons and as many as 20 to 50 times more glial cells supporting the functions of these neurons.

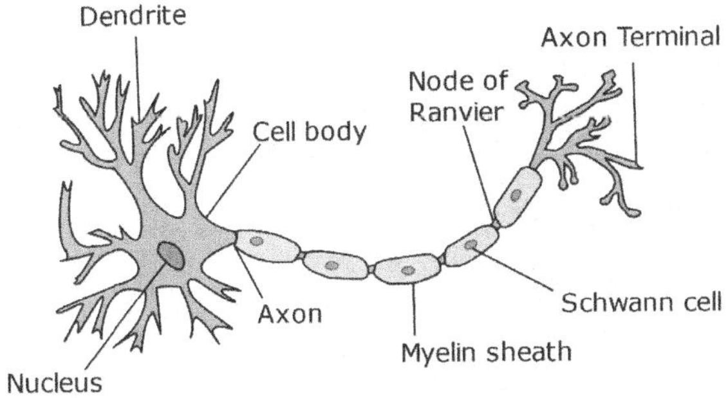

Fig. 2.1: Diagram of a Typical Neuron

Dendrites are an arbor of receptacles around the cell body of a neuron that help receive information from other neurons. Each neuron has a single long fiber called an axon. At the tip of the axon is a growth cone that serves to guide the axon to targeted brain regions. Once the axon reaches the target site, synapses, or points of connection, form between the axon and the target.

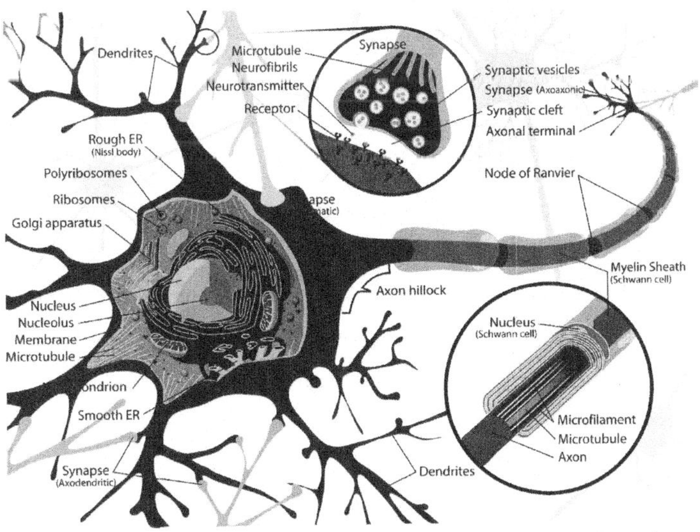

Fig. 2.2: Diagram showing a 3d cut section of a neuron cell with synaptic connections to other neurons and myelin sheath highlighted.

Bunches of neurons together form the nervous system. Neuronal axons can range in length from less than a millimeter to a meter or more. The longest axons run down from the brain into the spinal cord and beyond.

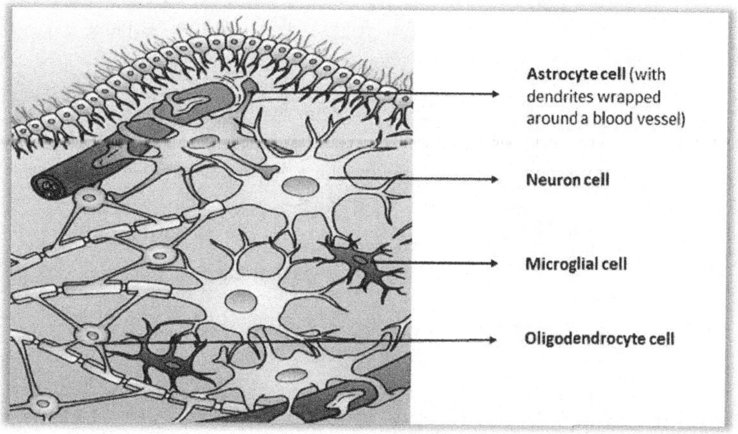

Fig. 2.3: Illustration of the three different types of glial cells found in the central nervous system.

In the brain, neurons with long axons are encased in a fatty substance called myelin, which is what makes the brain's white matter white. Myelin acts as insulation for axons, helping signals transmit over longer distances. For this reason, myelin is mostly found in neurons that connect different brain regions, rather than in the neurons whose axons remain in a local region. An axon typically develops side branches called axon collaterals, so that one neuron can send information to several others. These collaterals, just like the roots of a tree, split into smaller extensions called terminal branches. Each of these has a synaptic terminal on the tip. Groups of axons are organized into tracts by glial cells which help connect areas of the grey matter to other areas, whose coordination is necessary to carry out brain functionality. Every aspect of this intricate web of neurons and supporting cells helps conduct important brain functionality. Damage to any of these fine structures can result in loss of connectivity between areas and impact one or more aspects of brain functionality depending on the area of damage.

Different parts of the brain that serve different functions use this network of connections to communicate with other functional parts of the brain. Working together these neuron networks coordinate between our various senses and regulate our responses to sensory inputs that include visual recognition, imagination, auditory senses, olfactory (smell), gustatory (taste) and the rest of about 20 senses known to modern science including equilibrioception (sense of balance), proprioception (awareness of the position of limbs and body), vestibular sense (ability to feel velocity), thermoception (ability to feel temperature), kinaesthesia (sense of movement), and chronoception (sensing passage of time). It is believed some ancient civilizations were aware of 300+ senses, so science has a lot to discover yet about the working of the neurons in the brain and the different senses they support and coordinate. Clearly the relationship between these mostly unknown senses and the neurons involved ought to be pretty complex and hard to decipher.

Grey versus White Matter

The cerebral cortex, the outer folded body of the brain, shows two distinct areas of coloration when sectioned out: the grey matter and the white matter. While grey matter is more pinkish-white in reality, grey is the name that has stuck. The difference in coloration is due to the difference in composition of the two areas. The grey matter is mostly composed of the body of the neurons, the dendrites, synapses and axon terminals, while the white matter is composed of the myelin-coated axon parts of the neurons.

The grey matter is richly supplied with blood vessels due to the enormous needs for oxygen and energy to carry out the brain's mental processes. This makes the grey matter more susceptible to hypoxic damage under conditions of oxygen deprivation, an event we discuss in chapter 5. Grey matter in the various folded areas of the brain is connected to grey matter in other areas of the brain through the axons of the neurons in them. These distinct connected grey matter areas work together in groups with every group enabling a certain higher-order function. This coordination of the various grey matter areas is important to higher-order functions of the brain such as memory, language, imagination, critical thinking, etc., and damage to this area has corresponding impacts on these cognitive capabilities.

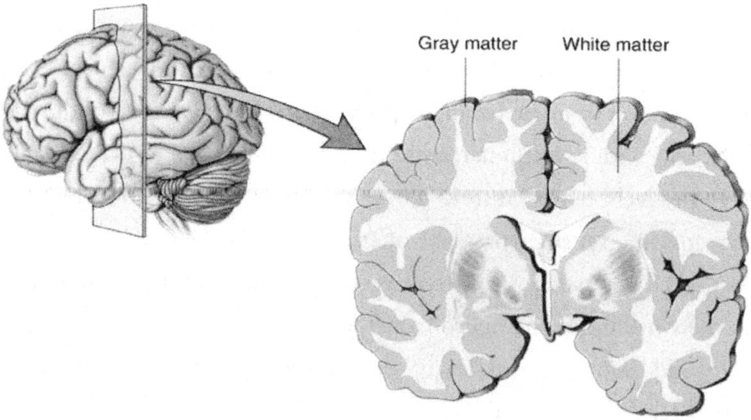

Fig. 2.4: Grey vs White matter.

White matter is made of the long axon fiber of the neurons in the grey matter. These axons are coated with myelin, a mixture of proteins and fatty lipids, that helps conduct nerve signals and protect the axons. This fatty myelin is also the reason for the whiter coloration of the white matter. Nerve signals travel between different parts of the grey matter and up and down the spinal cord through groups of axons called tracts that connect them. Damage to the white matter in the brain can result in severing of these connections and leads to loss of coordination between grey matter areas that is crucial to a lot of higher-order functioning. In case of white matter damage, the resulting inability for various areas of grey matter to coordinate might explain why individuals on the autism spectrum lack broad interests and tend to favor narrow areas of interest. The lack of conversational skills in autism might also be a result of this, since conversation involves interpreting language, understanding verbal and nonverbal cues, memory recall and correlation, which requires using different parts of the brain all at once in real-time.

Myelination

Most of the development in the early stages of childhood brain development involves the creation of pathways between the various functional areas of the brain and the myelination of the neurons connecting these pathways. Structural maturing of individual brain regions and their connecting pathways is required for the successful development of cognitive, motor, and sensory functions. Such growth is eventually responsible for a smooth flow of neural impulses throughout the brain, which allows for information to be integrated across the many spatially segregated brain regions involved in these functions. The speed of neural transmission is an important factor, and this depends not only on the junctions between nerve cells called synapses but also on the structural properties of the connecting axon fibers. Critical axon structural properties include their diameters and the thickness of the special myelin insulation around many fibers. Large groups of myelinated axons, which connect various regions in the brain, appear visibly as "white matter." Axon tracts of the major pathways carrying impulses in the human brain continue to develop throughout childhood and adolescence.

The nerve impulse is a rapid propagating wave, approximately 1 millisecond in duration, of chemical depolarization followed by repolarization. While it is really a transmission of a chemical impulse or action potential, the neuron axon can be thought of as behaving as an electrical transmission line whose cross-sectional diameter is an important determinant of impulse propagation velocity. The larger the diameter, the greater the velocity of propagation. Correspondingly, the myelin sheath that surrounds certain types of axons is a periodically interrupted type of electrical insulation, and its presence represents a substantial increase in pulse propagation velocity over that of a bare axon of the same diameter. The breaks in the myelin insulation, called the "Nodes of Ranvier," represent ion rich areas that help bolster the speed and strength of the nerve impulse.

Myelination is thus a major aspect of the workings of neural circuits. Any disturbance in the myelination process due to certain genetic conditions is known to lead to neurological disorders grouped under the heading of leukodystrophy. Such disorders are rare but have a genetic etiology. Unlike autism, which is a childhood developmental disorder, leukodystrophy conditions can be triggered even at later ages and is identifiable in an MRI scan. As rare as they are, at least 23 different known subdivisions of leukodystrophy are known to exist. Multiple Sclerosis (MS) is another disease that results from de-myelination of nerve cells. MS is caused by the body's immune system attacking the myelin sheath, causing damage to the brain and the rest of the nervous system. All this goes to prove the importance of myelination in the functioning of the brain and central nervous system – and how damage to these areas could have significant neurological impacts.

Neuron Fiber Lengths

Neurons make up most of the cerebellum (lower lobes at the rear of the head close to the neck) and cerebrum (2 upper lobes) of the brain. They are of different types and varying lengths and are often a source of confusion when trying to understand brain morphology and functioning. Not all neurons in the brain have the typical structure of a cell body with a very elongated axon attached. About 80 percent of brain neurons have very short axons, almost not distinguishable from the dendrites. The rest of the 20 percent have longer axons of varying lengths and traverse across the brain connecting its many regions. These longer axon neurons are also the ones whose axons are protected by the myelin sheath so they can transmit the action potentials across without loss.

Longer neurons are of particular interest due to the possibility of damage and a lesser possibility for recovery from damage. It is common knowledge that damage to the neurons of the spinal cord is the cause for paralysis in case of injury below the point of damage and such damage is mostly irrecoverable. The longest neurons in the body are those that run from the base of the spinal cord to the toes and are about one meter long. Those that run from the brain into the spinal column are also known to be very long. How long are the longest neurons in the brain? The answer depends on the location and function of the neuron within the brain, with lengths reaching the entire width of the two lobes of the cerebrum connecting areas in the left lobe of the cerebrum to areas in the right lobe for example. Neurons that project from the brain down into the spinal column are also very long.

Analysis of data from the neuromorpho.org database compared to the results of other scientists has allowed researchers from aiimpacts.org to arrive at the following conclusions for various neuron lengths in the cerebral and cerebellar brain areas:

Table 2.1 – Neuron lengths by location and type

Connection type	Average length per neuron (mm)	Contributing neuron types	Sources of evidence
Cerebellar, short-range	5.7 - 6.1	Granule (~70%), stellate, basket, Purkinje, Golgi	Morphometry
Cerebral, short-range	14 - 20	Pyramidal (2/3rds to 85%), stellate	Fiber density in rats, morphometry
Cerebral, long-range	100	Pyramidal	Width of corpus callosum, relationship between brain volume and global connection length

Source: aiimpacts.org

As seen in Fig. 2.5, the corpus callosum and anterior commissure are the neuron fiber bridges that connect the two hemispheres of the cerebrum. The projection fibers that constitute the internal capsule are neuron axon connections that connect the outer cerebral areas to the basal ganglia. The basal ganglia are the motor control centers which help decide between the various responses to the sensory inputs received.

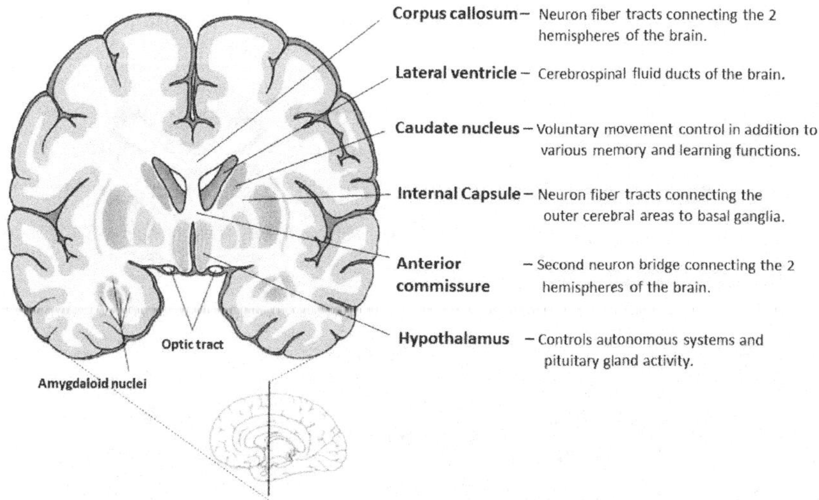

Fig. 2.5: Coronal or frontal cross-section of the brain, showing important neuron fiber connections of the brain and white matter functional control areas.

Traumatic damage to some of these longer length fibers in the deep brain areas can happen during birth as a baby's head makes its way out of its mother. For example, damage to the caudate nucleus is reported to result in loss of motivation, obsessive-compulsive disorder and hyperactivity. Even the smallest of scars in the brain can lead to negative outcomes for those babies impacted by trauma. Sometimes the damage is so small that it not visible in MRI images, due to the limitation of about one millimeter resolution of MRI, but the adverse effect on development is very real. This rarity of MRI visible scarring has also been one reason the traumatic brain injury at birth has remained hidden in most instances of autism.

Neurotransmitters

Neurons are connected to each other by special junctions called synapses. The synaptic junction allows action potentials to be transmitted to the target neuron using chemical molecules called neurotransmitters. The key concept to the synapse is that the neurons don't touch each other but have a very small gap between them called the synaptic cleft. Communication across this synaptic cleft is controlled by these neurotransmitters. Every neuron is connected to about 1000 other neurons. So, for the 100 billion neurons in the brain, there are about 100 trillion synaptic junctions.

Recall that proteins are made up of long chains of basic building blocks called amino acids. Neuro transmitters are singular amino acids or amino acid like compounds which convert the action potentials at the synaptic junctions between neurons into chemical messages triggering specific actions in the next connected neurons. The influence of the neurotransmitter on the next neuron or muscle could be excitation, inhibition or moderation.

Neurotransmitters are synthesized by the neurons at the axon terminals and stored in synaptic vesicles at these synaptic junctions. They are released into the synaptic cleft when an action potential is received from the axon and work to excite, inhibit or modulate receptors on the muscle or neuron they are connected to. They often trigger a response in the post synaptic cell. The synapses and neurotransmitters ensure that signals travel in only one direction. They also allow for signals to be amplified or moderated, and working with other neurons they this can result in amplification or reduction of impulses. Overall this can allow for complex signaling and a wide array of behaviors to emerge.

While there are hundreds of neurotransmitters in the brain, there exist 6 major neurotransmitters. The table below shows main neurotransmitters their functions and the results of their deficiency or excess.

Table 2.2: The Main Neurotransmitters

Neuro-transmitter & Effect	**Function**	**Deficiency**	**Excess**
Glutamate (Excitatory)	Primary excitatory neurotransmitter of the nervous system, controls all behaviors, learning, memory	Insomnia, concentration problems, mental exhaustion, low energy	Hyperactivity, anxiety, pain amplification, Neuron damage

Dopamine (Excitatory or Inhibitory)	Neuromodulator involved in voluntary movement, pleasure, motivation, learning	Parkinson's Disease (initiating movement), depression	Schizophrenia
GABA (Inhibitory)	Major inhibitory neurotransmitter, controls all behaviors including anxiety, motor control	Huntington's Disease (destruction of GABA neurons), loss of motor control	Inhibits brain function
Serotonin (Inhibitory)	sleep, memory, mood, pleasure	Depression	Prozac works by blocking re-uptake of serotonin
Endorphins (Inhibitory)	Inhibiting pain	Hyper-sensitivity to pain	Insensitivity to pain
Acetyl - Choline (Excitatory)	Muscular movement, memory, cognitive function	Alzheimer's Disease, paralysis (botulism)	Convulsions

As you can see these neurotransmitters can have opposite effects of excitation and inhibition. Some of the commonly paired neurotransmitters include Serotonin and Melatonin, Dopamine and Norepinephrine, and GABA and Glutamate. Additionally, neurons allow for co-transmission of more than one neurotransmitter from a single neuron. This makes it possible for even more complex behaviors to emerge.

Once the neurotransmitter has acted on the receptacles of the subsequent cell, it needs to be quickly eliminated so the next cycle of excitation or inhibition can begin. The brain has very efficient methods to perform this elimination including diffusion out of the synaptic cleft followed by absorption by glial cells, enzymatic breakdown by astrocytes, and re-uptake or re-absorption of the neuro transmitter back into the pre-synaptic neuron.

Thus, the complex biochemistry of these small amino acid molecules called neurotransmitters plays a very critical role in the functioning of the brain.

A significant portion of medications for autism and psychiatric disorders involve medication that tries to address supposed deficiency or excess of neurotransmitters. They are sometimes useful but outcomes very unpredictable due to the complex nature of neurotransmitter interactions. Individual reactions to specific medication need to be very carefully monitored if such a medical path is pursued.

Key Take Away: We have looked into the functioning of neurons, their physical structures, chemical and electro chemical signaling, using neurotransmitters. We considered the different excitatory and inhibitory effects of neurotransmitters. Other components of the brain physical structure like grey and white matter, myelin, all play very critical roles in the proper functioning of the brain.

CHAPTER 3

The Protected Brain

This chapter examines common protective bodily functions which tend to be performed slightly differently in the brain, compared to how they are performed in the rest of the body. This is due to the special status of the brain as the most important organ of the body. Just like understanding the development of the human brain is essential for understanding the afflictions that can come about as a result of disturbance or damage to any of these very delicate and complex mechanisms, it's important for parents of children on the spectrum to grasp how damage to brain function can impact their child – and why the brain by design is unlikely to repair itself as fast as (or as much as) other parts of the human body.

The Blood Brain Barrier

The blood brain barrier is a specialized protection mechanism around the blood vessels supplying the brain that regulates the flow of nutrients to and from the brain. This affords the brain an additional layer of protection from an infectious disease or other metabolic toxins. Inflammation in the tight cranial space can lead to neuron damage, and so this mechanism limits brain inflammation by limiting both disease vectors and immune cells from entering brain tissue. Additionally, being a very active environment, the brain needs a lot of efficient, active transport of nutrients (including removal of waste products). The blood brain barrier is designed to do just that with specialized layers of endothelial cells, astrocytes (support neurons) and pericytes (substrate cells).

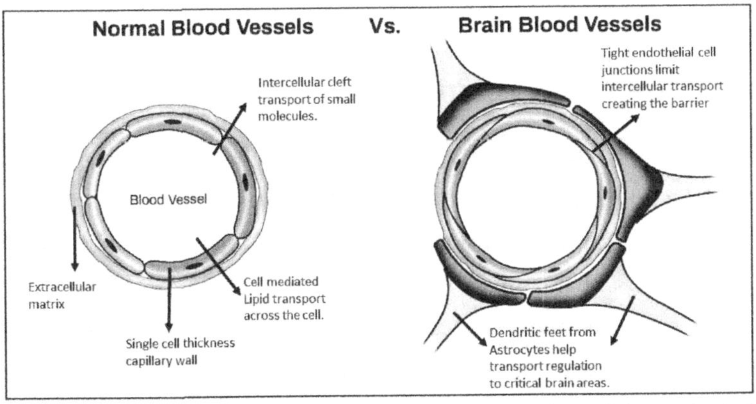

Fig. 3.1: Blood Brain barrier. Notice the tight overlap of the endothelial cells (red) in the brain blood vessel (right) creating the barrier. The astrocyte feet also modulate the uptake of molecules into the brain across the barrier.

Endothelial cells make up the walls of all blood vessels in the body. However, the endothelial cells in the blood vessels of the brain are special as they are bound together with tightly overlapping cell walls called "tight junctions." These tight junctions only allow for very selective transport of molecules between the cell walls of adjacent cells. The endothelial cells are wrapped by a layer of cells called pericytes that typically bind them to the surrounding tissue. However, inside the brain these are bound to specialized astrocyte neurons that excel at controlling the transport of nutrients into and out of the brain tissue which is made up of glial cells and other types of neurons.

As I discuss in detail in later chapters, the blood brain barrier is a very important and distinguishing feature of the brain because while it is mainly beneficial due to the protection it provides from external threats, it also presents a challenge when treating brain conditions. Medicines meant to treat the brain may not make it past the blood brain barrier. Alzheimer's is suspected to be caused by plaque buildup in the brain from the inability to drain certain substances out across the blood brain barrier. Brain injury can compromise this barrier and this limits the ability of the central nervous system to heal itself. This is one of the reasons brain damage is actually permanent in many cases.

Brain Immune Privilege

Certain organs in the human body seem to have evolved such that they are able to tolerate the presence of foreign substances (antigens) inside of them. The presence of antigens inside of these organs doesn't evoke the typical immune response that happens everywhere else in the body. This includes the eyes, testicles, placenta and fetus and the central nervous system (CNS). These organs are said to have immune privilege, as typical immune responses in these areas could be more damaging than useful; hence evolution seems to have taken a path to allow these organs to stay outside the bounds of the immune system.

Originally the entire brain was thought to be immune privileged by virtue of the blood brain barrier. It's now been proven that the brain immune privilege is limited to certain parts of the central nervous system. The brain has a different type of immune response from the typical peripheral immune system response. While immune cells are typically not allowed past the blood brain barrier, at times of need, there is complex signaling that allows certain special immune cells across the perineal, endothelial and astrocyte layers and enter the brain vascular regions.

The exact working of the immune system is not fully understood and still a subject of intense research. There is an enormous amount of complex physiology involved in the inner working of the immune system in higher-order vertebrates. The major organs involved in immunity include adenoids, bone marrow, lymphatic system, spleen, thymus and tonsils. In very simplified terms, there are 2 layers of immunity. The innate immune system consists of the skin, stomach acid, enzymes found in tears and skin oils, mucus and the cough reflex and is the first line of defense. Innate immunity also includes cell mediated defenses like regular white blood cells that engulf and digest invading pathogens and the lymphatic system that creates and maintains these white blood cells. The key distinguishing feature of innate immunity is the fact that this defense is not against specific pathogens, but directed toward a broad range of common antigen (foreign body) characteristics learned over the course of evolution. The innate system also has ways of triggering the adaptive immune system when unable to take care of the situation by itself.

Active immunity is the second line of defense against the pathogens of the outside world. The active part of innate immunity involves two kinds of immune patrol cells, macrophages and dendrites found at the tissue level. Macrophages are cells that engulf and digest identified pathogens and dendritic cells are cells that attach to the pathogen and inactivate them or return to the lymphatic system with a template of how the foreign body can be attached to. When innate immunity fails to stop invading microbes, the adaptive immune system is called into action by dendrites returning with templates for attacking the foreign body or pathogen (antigen). The adaptive immune system is a more directed response to the specific antigen consisting of white blood cells created in the lymphatic system. At this point the body tissue level response turns into a blood cell, whole body response. The white blood cells of the adaptive immune system are of 2 types the B-cells (plasma cells) and the T-cells. The B-cells have a large number of receptors that bind to specific antigens. Once bound the B-cells return to the lymphatic system to create more copies of themselves. These then go out to bind more antigens and present them to T-cells for further action. They do this presentation by introducing binding proteins called antibodies into the bloodstream so more antigens can be identified and removed. The subtype of T-cells called the killer T-cells are responsible for removing the antigen cells from the system using an antibody dependent digestion mechanism. They are generated in the lymphatic system in response to the antigens bound by the B-cells. This dual response with directed antigen response, from the lymphatic system is the signature response of the adaptive immune system.

This very simplified version of the entire immune response system is captured in fig 3.2:

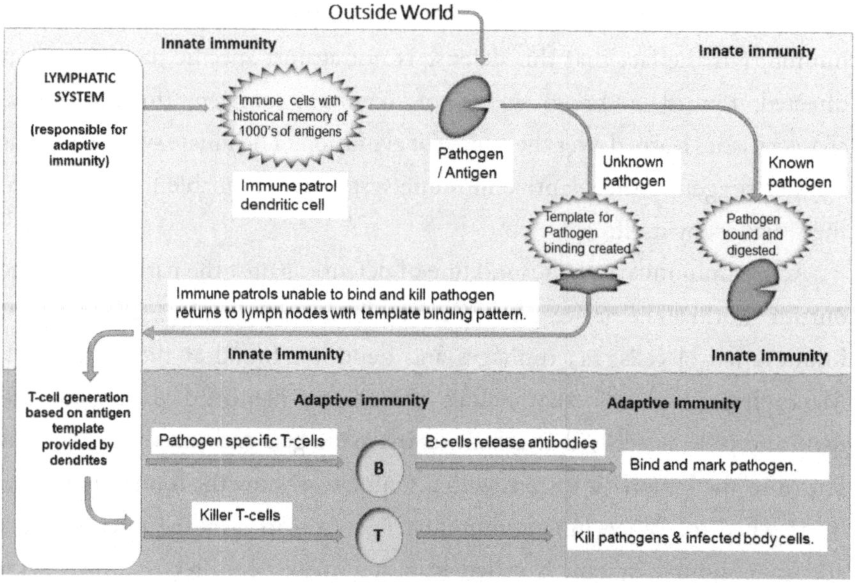

Fig. 3.2: An oversimplified view of the immune system showing innate immunity, adaptive immunity and their relationship to the lymphatic system.

The immune system is very complex including certain other proteins like cytokines, interferons that I have not discussed here for the sake of simplicity. The actual roles of certain organs, cell types and proteins in the immune system are still being discovered and identified.

ACAID Phenomenon

The blood brain barrier seems to offer a separation of the brain immunity from the rest of the immune system. But recent discoveries suggest there is still interaction of the brain with the B-cells of the immune system. The brain lacks a lymphatic system, but there is evidence of the brain tissue draining antigens into the cervical lymphatic nodes found in the neck. The immune system response to antigens coming from the brain is heavily skewed toward the B-cell response. This avoidance of the T-cell response in the brain is what makes the brain immune response different from elsewhere in the body. In fact, re-introduction of the same antigen in other parts of the body after a previous encounter in the brain also leads to the lack of a T-cell immune response. Originally observed in the anterior chambers of the eyes, this phenomenon is called ACAID (Anterior Chamber-Associated Immune Deviation).

ACAID is a phenomenon of immune suppression that is known to exist in many mammals (starting with mice all the way to higher primates). It was first noticed when corneal transplants were observed to be the least rejected transplant by the immune system. The eyes are an immune privileged organ protected from the immune system by the blood-retinal barrier. This is because evolution has over long periods identified that a typical immune response in some areas of the body is actually more detrimental than useful. It is observed that introducing an antigen (foreign particle) directly into the anterior chamber of the eye results in immune suppression for that antigen in the rest of the body. Research is underway to look into ways of taking advantage of this phenomenon to treat autoimmune diseases, organ system transplant rejects, etc. These are the positive aspects of the ACAID phenomenon. However, if foreign organisms directly invade these immune privileged organs (perhaps via damaged areas in the barriers that typically protect them), then they might end up obtaining immune suppression in the rest of the body, leading to chronic illnesses that are hard to treat.

The white matter of the brain is one area where the blood brain barrier creates a different immune environment. Not much is known about ACAID immune suppression in the case of white matter, as human testing for something like this is not possible. Nevertheless, this phenomenon of ACAID combined with a compromised blood brain barrier in subjects with brain damage might actually play a role in why viruses might survive in some other parts of the nervous system and continue a chronic, low grade, long term adverse impact on the affected individual. In the rest of the brain tissue, glial cells form part of the regular response in addition to the innate immunity afforded by the blood brain barrier, and there is interaction between the neuronal and immune systems (particularly at the cellular level where the glial cells aid in the removal of antigens, astrocytes aid in the transport of antibodies and B-cells across the blood brain barrier). There is also evidence of B-cell response, but the inflammatory T-cell response has not been found in the brain. But the white matter of the brain is the only area that is totally immune privileged – and therefore cordoned off from easily repairing itself – or ridding itself of viruses as well.

Neuroplasticity

Neuroplasticity is a word coined in recent times to describe how the brain and its connections are constantly changing as a result of external stimulus and environmental input. Practicing certain activities is known to reinforce the corresponding areas of the brain where the functionality resides. We all recognize the saying "practice makes a man perfect." Apparently, the reasoning behind it is neuroplasticity – the re-wiring of your brain to do that task perfectly. The proportion of grey matter and synapses in the various areas of the brain may also change over time corresponding to the activities the individual predominantly engages in.

Another area where neuroplasticity helps is the possibility of recovery from injury. Studies on monkeys and mice have demonstrated the ability of their brains to recover from injury over long periods of time. It's been demonstrated they develop the same functionality of the damaged area of the brain in areas adjoining the damage with adequate external stimulation. This ability of healthy individuals to recover from damage seems to occur at all ages except in individuals past the age of 70 or so, where genetic damage becomes more significant and reduces the chances of neuroplastic recovery.

However, the ability of the brain to heal itself due to neuroplasticity is also misunderstood. It is far from the truth that the brain can heal itself. The extent of damage and location of the damage plays an important role in any possibility of recovery from injury. Scar tissue occurring from damage is known to inhibit regrowth of neurons in the scar and surrounding areas. It represents a major challenge to recovery. In cases of significant scarring damage to the brain or other parts of the CNS, the damage is usually permanent given our current knowledge of brain healing and treatment options. Also, if the ability to recognize and respond to external stimuli is lost as a result of damage, the ability to recover is lost as well.

Traumatic Brain Injury (TBI)

The brain is the most fragile of all living tissue, with a very fine and intricate neuron structure in the white matter. It is also the most sensitive to changes in oxygen levels and oxygen supply. The brain requires about 15% of cardiovascular output in adults – a number known to be even higher in children. The system is so critical that the 2 main arteries supplying the anterior brain and the posterior brain and brain stem back each other up, through the bilateral communicating posterior arteries.

Hypoxic brain injury happens when there is an insufficient supply of oxygen reaching the brain tissue. This results in neurons dying out from lack of oxygen causing permanent brain damage. It is called cerebral hypoxia or cerebral anoxia. Hypoxia can happen to a fetus at or near birth due to many causes, but it is mostly related to the constriction of blood supply to the fetal brain. Any constriction of blood supply to any part of the body is called ischemia. So, it is not surprising that brain injury at birth is referred to as hypoxic-ischemic injury. It can be caused by any one of many causes that can reduce blood flow to the fetal brain.

Many natural pregnancy complications related to birth cause hypoxic brain damage, resulting in neurodevelopmental disorders ranging from cerebral palsy to high functioning autism. Hypoxia or loss of oxygen to the fetal brain can be due to low maternal blood pressure, placental detachment, inflammatory disease in maternal pelvis constricting blood flow to placenta, and blood clotting complications. Other birth complications like placenta placement, cord compression, nuchal cord and other less common malformations of blood vessels can lead to hypoxia as well. And it is also the case that extensive and prolonged contractions – particularly when artificially induced at a high frequency– can create pressures that choke off blood supply to the fine capillaries of the brain.

Extensive damage to brain tissue can also occur as a result of the pressure on the brain violent contractions can cause:

- Pressure can cause physical trauma to the long axons of the neurons in the white matter, which are only a few nanometers thick and fragile enough to break.
- Excess pulsating pressure at the same points for a long time can also cause the cell bodies of the neurons and glial cells to get damaged from the physical trauma.

When the brain is injured due to external trauma to a point where its functioning is adversely impacted, it is said to have sustained a TBI.

Of particular interest in the case of autism is the possibility of traumatic brain injury at birth. There are three causes of TBI at birth:

- Large head of the baby or small maternal hip (cephalopelvic disproportion);
- Rupture of prematurely fused skull plates (craniosynostosis) driving fragments into the brain during birth;
- Force of contractions hammering the fetal head against a non-effacing, non-compliant cervix.

The damage from TBI can be to the grey matter, the white matter or both. As neuron bodies in the grey matter have their axons projecting into the white matter and damage to any part of the neuron results in that neuron dying out, the damage permeates the entire brain regardless of where the damage originates.

At this point it should be clear that despite neuroplasticity, damage to the brain could be particularly debilitating due to the body's inability (by design) to repair those regions. The most serious problem with brain trauma at birth is that it can be distributed over large areas of the brain. Such brain injury results in neurodevelopmental delays and disorders that I will argue later are often classified as autism. Additionally, scar tissue is associated with seizures and complications with motor control while preventing the regrowth of neurons making such damage permanent.

Gut Brain Axis - The second brain

The human gut has up to 200 million neurons which form an important part of the digestive system. This is so important that it is considered its own separate part of the nervous system called the 'enteric nervous system'. The guts form the basis of the fight vs flight response, so the neurons of the guts might actually be involved in the 'gut instincts' that people frequently talk about!

The gut–brain axis is the biochemical signaling that takes place between the gastrointestinal tract and the central nervous system. Key to this signaling is the chemicals created by the microbes of the gut. Hence the term has now been expanded to be called the microbiome gut-brain axis. Research interest was sparked by a 2004 paper from Kyushu university Japan showed mice with a sterile gut environment were more prone to higher stress responses.[3.2] Subsequent microbiome research, mostly done in mice has shown the importance of the gut biome in the normal functioning of the brain.

Exciting new relationships about the gut flora or microbiome and a person's mental and physical health are being discovered fairly frequently these days. This is leading to more acceptance and common use of probiotics and pre-biotics as health supplements in our times. The consumption of foods for their probiotic value like yoghurts and fermented foods are known in many cultures around the world.

Gut dysbiosis is a common symptom in many individuals suffering from autism. Numerous autism treatments also focus on alleviating the gut-dysbiosis in autism as a pathway to helping alleviate autism symptoms. There is also evidence that victims of traumatic brain injury invariably end up with gut problems and there are reasons for why this is the case. I will reveal the reasons and true causes for these relationships in the upcoming chapters.

Key Take Away: The immune process works differently in the brain than elsewhere in the body due to the low tolerance for swelling and also due to the protection of the blood brain barrier. This brain immune privilege, combined with the ACAID phenomenon, might actually be detrimental if invading antigens gain entry into the brain through compromised blood brain barrier areas as in instances of traumatic brain injury. And traumatic brain injury can be brought about by the force of contractions hammering the brain against a non-compliant cervix.

Part II

Autism and TBI at Birth

TBI is a possibility during childbirth and has historically been the major cause of mental retardation throughout history. A baby making its way out of the mother head first has always been a brain damage risk. The natural birth process has developed so as to minimize this risk. When this natural process is interfered with, it is sometimes a recipe for disaster. In short, your brain health and intelligence are in large part determined by how you were born – literally.

CHAPTER 4

The Perils of Childbirth

I am a parent of two boys on the autism spectrum and have been in the unique position that only autism parents find themselves in: of having front row seats to the events that unfolded in the lives of their kids, right from pregnancy and birth to their diagnosis and treatments. To explain why the causes of autism are not primarily genetic but instead are the result of birth complications, which sometimes include medical malpractice and drug injury during birth, I want to share with you some of the basics I needed to learn in order to look past the deceptions we have been fed. It takes a certain level of understanding of the many factors that impact autism to see the real reasons behind it, and the first and foremost of those is understanding the perils of childbirth.

Childbirth Then and Now

Throughout time, the birth of a child is a happy occasion for the entire family and a cause for celebration in all traditions of the world. The arrival of that bundle of joy changes the lives of a couple forever, propelling them forward to a new stage of life as parents – a stage which has the potential to fill the new parents' lives with joy and laughter and brings a certain meaning and purpose to life.

However, historically things have never been this simple. There were times not so long ago when women did not go to the hospital to have a baby. Childbirth was fraught with danger to the lives of both the mother and the child. A woman who survived her first pregnancy was thought to have been reborn. In peaceful societies of ancient times, where men were not dying in wars and conflicts, the male-female ratio was tilted heavily toward men simply due to the fact that a significant population of women died due to complications of childbirth. Even until a century ago, pregnancy was so much of a risk that entire traditions and customs about what happens if a mother dies in pregnancy were observed. One remarkable success of modern medicine has been the ability to reduce the risk of childbirth, to a point where pregnancy is considered a mostly secure and safe process, with very little or no risk to life at any point. A lot of this success can be attributed to the mastering of safe C-sections.

The C-section is one of the first surgeries of modern times. Originally, C-sections were a last-ditch effort to save the baby while the mother was left to die. With the arrival of modern medicine came the option of anesthesia and surgery where the mother was sewn back up and lived. In the last 50 years, the whole birth process has become much less of a risk after the development of the standard C-section delivery using modern surgical techniques.

All this begs the simple question: do animals die during pregnancy and birth? The answer is yes. The birth process can be a challenge for first-time mothers everywhere in the mammalian kingdom, although it depends a great deal on the size of the newborn relative to the mother and the length of the labor. Animals are known to go through pain during labor, although labor in animals is much shorter than in human beings. Certain species are known to lose 15-20% of first-time mothers to delivery-related deaths—a number similar for humans until the advent of modern times and advances in C-section delivery.

The Birth Process

Like the rest of the newborns in the mammalian world, human babies are born coming out of the birth canal headfirst (the exceptions to this are the water-dwelling mammals like dolphins and whales whose calves come out tail first, which helps prevent the possibility of drowning/suffocating as they get pushed out slowly). Nature and evolution have taken care to ensure that delivery does not hurt the baby in any way. The child is kept protected inside the amniotic sac surrounded by amniotic fluid inside of the mother's uterus to ensure its safety (think airbags protecting passengers in an automotive crash). Such is the level of protection provided to the baby by the natural design. Nature has perfected this process over hundreds of thousands of years making sure it is almost flawless. That is not to say breech births – where the baby comes out legs first – do not happen on some rare occasions both in human beings and in the animal world. But even in such cases techniques have been developed by traditional midwives to ensure the safe delivery of the baby.

The birth process is by itself a miracle of nature. But the safe protection of the baby also means the child has to break or pass through several barriers head first to arrive into this world. These barriers are:

- Amniotic sac and fluid-filled amnion;
- Cervix uterus wall; and
- Pelvic bones.

The schematic diagram below shows the "head-on" task for the baby to be born

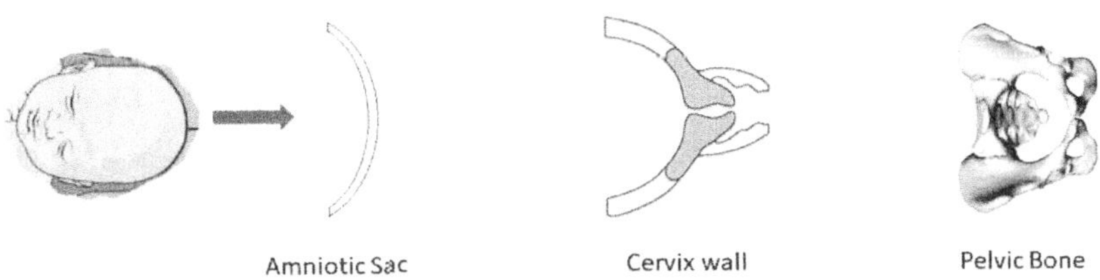

Fig. 4.1: The Three Barriers to Birth

Overcoming each of these barriers is challenging. The pressures the head is subjected to during the birth process can lead to brain injury depending on the size of the baby's head and the pliability of the cervix wall. It should be noted that the birth process has historically been the leading cause of brain damage and mental retardation in infants and in extreme cases can even result in the death of a baby. Loss of oxygen to the brain from reduced blood flows during contractions is another danger that has to be navigated in a successful birth.

While fraught with dangers, the labor process itself (as shown in Fig. 1.2) occurs naturally in what is under normal circumstances a positive feedback loop:

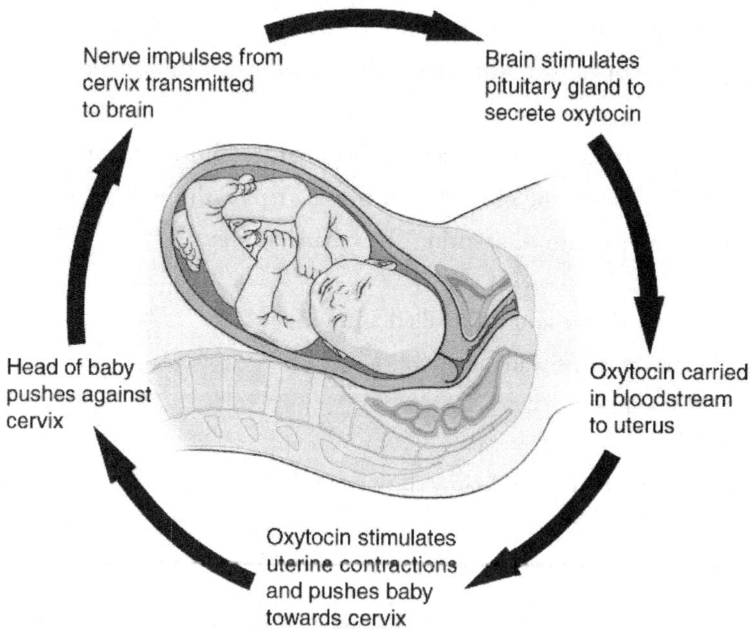

Fig. 4.2: The positive feedback loop in birth.[4.2]

The first contractions of labor squeeze the upper end of the uterus called the fundus, pushing the baby toward the cervix at the lower end of the uterus. The cervix contains sensitive nerve cells that monitor the degree of stretching. These nerve cells send messages to the brain, which in turn causes the pituitary gland at the base of the brain to release the hormone oxytocin into the bloodstream. Oxytocin causes stronger contractions of the smooth muscles in the uterus, pushing the baby further down the birth canal. This causes even greater stretching of the cervix. The cycle of stretching, oxytocin release, and increasingly more forceful contractions leads to the effacement of the cervix and the baby is released into the birth canal, ready to be pushed out.

However, this positive feedback process doesn't happen as smoothly as this description might make it seem. Enormous changes in the mother's body are required to expel the baby at the end of pregnancy. And the events of childbirth, once begun, must progress rapidly to a conclusion or the life of the mother and the baby are at risk. Unfortunately, most of the literature out there does not elaborate on these risks and the potential for lifelong consequences (particularly for the baby) when this rapid progress requirement is not observed religiously.

The Three Stages of Labor

The process of labor is divided into three stages in the medical literature.

First stage: This is the longest-lasting and arguably most significant stage of labor. It lasts from the time of onset until the cervix is completely dilated to 10 cm. Medical science divides the significant activities that happen during this stage into 3 phases.

- **Phase 1** (Early Labor): Onset of labor until the cervix is dilated to 3 cm.
- **Phase 2** (Active Labor): From 3 cm dilation until the cervix is dilated to 7 cm.
- **Phase 3** (Transition): From 7 cm dilation until the cervix is fully dilated to 10 cm.

This stage also includes the rupture of the amniotic sac in the initial stages of labor and the baby making its way past the first 2 barriers that keep it safe inside the mother during pregnancy.

Second stage: The period after the cervix is dilated to 10 cm until the baby is delivered. This is the pushing stage where the baby needs to be pushed out through the birth canal past the pelvic bone structure. This can result in the "crowning" of the head where the disjointed bones of the baby's skull deform to take a conical shape allowing easier passage through the birth canal.

Third stage: Delivery of the placenta, which includes the whole placenta organ along with the umbilical cord and the amniotic sac.

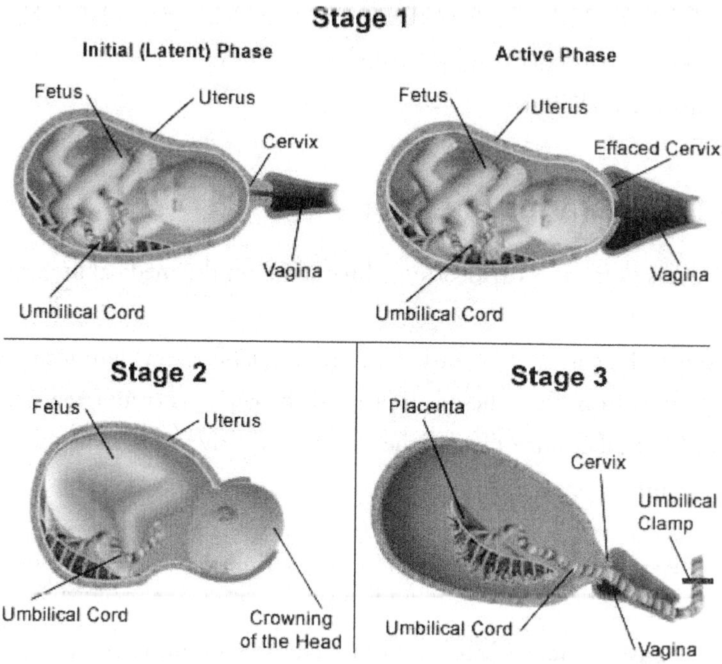

Fig. 4.3: Diagrammatic description of the 3 stages of labor.

While these stages are a generalization, every birth story is unique and no two deliveries go exactly the same. But in the vast majority of cases this seemingly enormous task of breaking and passing through these three barriers head first happens without a glitch. The water breaks, causing the uterine contractions to go faster, which in turn causes the thinning out and disappearance of the cervix wall. The baby can then be pushed out by the mother through the pelvic bone structure to be born into this world.

But this process has always been fraught with danger and has historically been the biggest source of maternal and fetal deaths. The amniotic sac is a tough membrane that typically remains intact until the end of the pregnancy, but there is the danger that it could break prematurely. Once the amniotic sac is broken it's advisable to deliver the baby sooner than later. The uterus cervix wall may also refuse to dilate and efface. This is where induced labor comes into the picture. A third complication can arise involving the bones of the mother's pelvis (including the pubic bone), which are designed to give to allow a baby to pass through. But sometimes the baby can remain stuck in the tight opening of the birth canal, necessitating a C-section. Finally, the baby's head is capable of deforming to take on a conical shape as it makes its way through the birth canal. This is due to flexible skull bones and disjointed skull plates, both of which can be damaged in delivery. In short, despite all these bodily adaptations in both the mother and the baby, birth remains a very challenging endeavor. The fact that most pregnancies go off without a hitch due to modern science means that these challenges are not given their due importance – obstacles that when not surmounted result in life-changing consequences for some newborns and their families.

Challenges for Mothers During Birth

Some of the major challenges for mothers include long labor times, large infant heads, and narrow maternal hips. Then there are these other rare complications that can happen where the mother's life can be in serious danger. Although considered rare, the occurrence of some of these complications has been on the rise. Typical complications have been overcome by modern medications and diagnosis methods, but some unanticipated issues like blood clotting and amniotic fluid embolism that are very hard to detect continue to pose a risk. Table 1.1 outlines some of these major risks.

Table 4.1: Risks to the Mother during Pregnancy

High Blood pressure	Pre-eclampsia is a pregnancy complication characterized by high blood pressure and signs of damage to other organ systems like the liver and kidneys. It's essentially a sign of the mother's organs working harder to keep the fetus supplied with blood and nutrients compensating for a weaker attachment of the placenta to the mother's uterus.
Placenta issues	Abnormal placement or growth of the placenta in the womb can pose a serious risk to the mother and baby at delivery. For example, placenta praevia is a condition where the placenta attaches on the lower end of the uterus where it can interfere with or even block the cervix preventing vaginal delivery. Other placental complications can occur when the placenta attaches itself at the site of a uterine scar from a previous C-section or fibroid removal. Such an attachment can be deeper than typical complicating birth and postnatal recovery.
Hemorrhage	Excessive blood loss during and after childbirth, where the site of placental attachment to the uterus leaves a gaping wound that requires quick closure. Uncontrolled blood loss might require the removal of the uterus (hysterectomy) to save the mother's life.
Blood Clots	Blood clots in the legs or lungs (also known as thromboembolic conditions) are a leading cause of illness associated with pregnancy and birth and can be life-threatening.
Sepsis	Sepsis is an infection of the blood that can develop before or after the baby has been delivered. Infections can be more severe in pregnancy, when women may be at particular risk of infection of the womb or birth canal and after delivery, with the site of placental attachment at particular risk for infection. This was a leading cause of maternal deaths before the advent of antibiotics.
Amniotic Fluid Embolism	A very rare complication of pregnancy in which amniotic fluid, fetal skin or other cells enter the woman's bloodstream and trigger an allergic reaction. Women with this condition may collapse suddenly during the birth of their baby and it often results in the death of the mother.

Challenges for the Baby During Birth

The list of things that can go wrong for a baby are actually longer and more common, and includes every risk to the mother (since the baby is completely dependent on its mother). Caregivers and doctors typically do not discuss many of these to newly expecting couples so as not to induce a lot of fear and anxiety during a time when they should be happy and excited. While arguably wise, doing so does not diminish any of these serious risks that are part of any pregnancy:

- Premature rupture of the amniotic sac;
- Premature placenta abruption (detachment from the uterus);
- Fetal Acidosis from placenta complications at birth;
- Premature contractions and birth;
- High blood pressure in mother;
- Gestational diabetes in mother;
- Umbilical cord around the neck;
- Kinked umbilical cord with fetal distress;
- Umbilical cord prolapse (cord stuck ahead of head during birth);
- Breech (baby does not turn to be head down); and
- No labor past 40 weeks (leading to cord ripening and detachment, meconium poisoning).

Every one of these conditions can result in life-threatening circumstances for the baby. Even with modern medical interventions babies are still lost to these complications, with the risk of death is still much higher to a baby than the mother. There are consequences of permanent brain damage to a baby from each of these risks, which can cause lifelong disabilities. As I will argue, the milder effects will end up being called Autism, ADD, ADHD and so on, though the link between the two has been de-emphasized.

It obviously takes exceptional skill and judgment on the part of an obstetrician to navigate through these enormous complexities. The US is credited with having the most advanced health care system in the world. For the birth process, this means a wider array of drugs, a wider array of interventions, and consequently a greater possibility of drug side effects and medical errors. The US also has the highest incidence of autism in the world. The correlation is no accident. Autism is considered a disease of the advanced western countries – the price of technological advancement, as the pundits like to put it. Contrast that with the lower or non-existent autism rates in countries where the only birth intervention outside of natural remedies is the C-section. The logical conclusion is that there's a correlation between birth practices and these autism statistics. Perhaps in many cases allowing the natural processes to take their course as in other countries is better than advanced interventions. We will unravel the reasons behind these statistics in the US, and explore solutions in the upcoming chapters of this book.

Standard of Hospital Delivery Care

Perhaps these findings about birth practices don't appear as shocking when you consider the fact that the standard of hospital delivery care in the US is the lowest in the industrialized world. This is as per the 2013 research report.[4.4] by savethechildren.org, an effort supported by the Bill and Melinda Gates Foundation, which ranked the US at 30 among 186 countries in the Birth Day risk index (which compares the first-day death rates among babies). The report in subsequent years since hasn't shown any significant improvement either. In the same report, another parameter to gauge how mothers fare, the annual mothers index, showed that the US had the highest maternal death rate in the industrialized world.

The US has a very technologically advanced but inconsistent medical system. For every expert physician there is another medical professional whose bungling of care, outright disdain for the patient and sheer incompetence – combined with bureaucracy and the practice of defensive medicine – has led to a system that's expensive and ineffective.

Some of the failures in the US can be attributed to factory-style hospitals trying to optimize their patient care and the doctor interaction times, medical insurance controlling care, but some are also attributable to doctors themselves practicing defensive medicine instead of learning the art of diagnosis by interviewing their patients. This results in doctors missing out on patients not falling into their standardized template of care. When care in the obstetrics front suffers, it points to systemic issues, particularly in the competence and training of OB/GYNs. Recent research shows that only about 2.3% of doctors are responsible for 38.9% of all medical malpractice claims and 94% of doctors had no claim against them[4.1]. We could expect these kinds of distributions might happen with OB/GYNs as well.

Key Take Away: Historically, brain damage at birth has been the single largest cause of mental retardation and lower brain functionality and IQ. Since a baby has to make its way out of the mother head first, physical brain damage at this stage is a very real possibility, and birth injury is the primary cause for a whole host of mental disabilities in children. But modern medical advice has sought to downplay the brain damage risk to a baby during birth as a consequence of past experience with birth injury litigations, resulting in focusing autism research onto genetic and other non-controversial areas.

CHAPTER 5

The Forgotten Story Rekindled by an MRI

There are a lot of things that I wish were different about our story, but perhaps things happen for a reason. These days all I ask for is a quiet regular life, but if things had been regular and quiet, I wouldn't have had anything to write about. Revisiting some of the old memories and incidents that have changed my life and that of my wife forever is never easy, and particularly those surrounding the birth of my kids are to this day very painful. But I have to tell our story so the unfortunate events which led to the current circumstances of our family will uncover a truth that would save thousands of other children and their families from falling prey to autism.

Delivery #1

My wife and I were in many respects just like everyone else in America – a classic story of a young couple caught up in pursuing the American dream and raising a family. We were both young, educated, and enjoying our own careers in software engineering. Two years after we got married, we decided to start a family. But pregnancy wasn't something that happened immediately; in fact, it didn't happen for quite a while. Finally, two years after our decision to start a family my wife was pregnant and we were ecstatic. The joy was just multiplied so much more when we finally found out after having waited so long. We were on our way to become the happiest parents around – or so we thought. Life had a lot more unexpected things in store for us. But if you could have seen us then you would not know what the future held. We were just happy, eager, expecting parents getting ready to welcome that new bundle of joy into our lives. Nothing could be more exciting and we spared no expense buying new baby stuff: a new crib, crib décor, car seat, and stroller. When the date drew nearer, we even upgraded one of our cars to a new minivan to bring the baby back home in.

We had a wonderful primary care doctor and my wife had picked a gynecologist – we'll call her Dr. S. – that our primary care practice recommended. She was a well-spoken pleasant doctor who did give us a lot of good advice on our pregnancy. She was obviously trying to do a good job in her profession. We had no reason to believe there was any incompetence on her part. But, years later, we recognize our pick of Dr. S. was a fateful decision that turned our lives upside down and something we'll regret for the rest of our lives.

The pregnancy itself was picture perfect. My wife had no morning sickness, no complications whatsoever, no diabetes, no blood pressure, no sickness or medication of any type. Dr. S. kept reiterating about how she would have to stay as active as possible and how staying active was very important to avoid the worst possible outcome in pregnancy: a C-section.

My wife continued to religiously take her prenatal vitamins, we ate fresh healthy homemade food, went out on walks and did everything the doctor asked us to do. As the due date approached, the doctor checked her out and said the baby was head down, ready to come. About a week before the due date there was still no cervical dilation and no sign of labor pains. Dr. S. at the point told us about the risks of going past the delivery date, including cord ripening and meconium being a risk. She also told us about how the baby might get bigger if we waited and then we'd be forced to have that dreaded C-section. She went on to introduce us to the concept of induced labor. She mentioned how it was a "safe and natural" method. She even explained how the pelvic bones are designed to expand to make way for the baby and how the baby's head could change shape to come out and how the brain was neuroplastic and recover from the changes. Dr. S. went ahead to schedule us for induction right on the due date, which we now know is something against the norm for a first pregnancy, where it is typical to wait 2 weeks rather than induce right on the projected due date.

We showed up at the hospital at 8:00 am on the morning of the appointed due date. She was checked for a bit and induction was started. It went on for about 6 hours but nothing seemed to be happening – she simply wasn't dilating. Around midafternoon the nurse came by to say we were not ready to have the baby yet and that we should go home now and schedule again in a week or so if nothing changes in the interim. At this point my wife's water hadn't been broken and nothing had changed. We didn't quite understand what was going on but decided to follow the medical advice and packed up and left the hospital. I believe the nurses had intervened at this point to educate the doctor about the 2-week waiting period in case of first pregnancy and possibly Pitocin contra-indication. I wish the nurses had explained better about the 2-week grace time over the 40 weeks in case of the first pregnancy, so it might have raised the alarm bells in our minds about the doctor. But they were not legally allowed to give medical advice.

We were asked to pick a second appointment, which we chose that same weekend about 5 days later. Nothing had changed in those 5 days; she was still not dilated. This time, however, they used a speculum to open the cervix and forcibly broke the placenta water (amniotomy) early on to try to get the dilation going. Induction went on and on for a total of 12 hours until our boy was born. Finally, we held that bundle of joy in our hands and were ecstatic.

As first-time immigrants used to a totally different tradition, we couldn't quite buy into the concept of starting out kids in day care at 6 weeks. We wanted to make sure our precious newborn had more time at home, so we requested that our parents who were visiting from overseas to see their new grandchild to extend their stay by a few months. We did everything imaginable to give him a healthy start – we even stayed away from processed foods. All signs were good: our baby was growing very rapidly and at the 100th percentile on some measures in those initial months.

Trouble Begins

But as he started to grow up, we noticed our boy was hyperactive and hard to keep up with. As a baby he was always restless and wriggling, and from the time he learned to walk he would run everywhere instead. Unlike other kids, his first words at 5 months old were not 'ma' or any of its variants; instead, his first word was 'car,' and his fascination with mechanical objects (over people) persisted as he grew up. We did not know what to make of these differences, but since he was hitting or exceeding all his milestones of growth we didn't think much of it, assuming instead that we had a precocious genius who was hard to keep up with. One other thing we noticed was his vision was a little poorer than it could be. He would look at things by putting them very close to his eyes. But they said he had to be four until he could be properly tested out and he would improve if we took him out to parks and tried to make him see longer distances, which we religiously did whenever we had a chance.

It wasn't until our older boy was age 3 when we finally reported back to the pediatrician that something was wrong. He was not talking like the other kids and wouldn't even answer simple yes-no questions (although he had quite a few words and was otherwise very active). He did fall sick occasionally, but he wasn't any worse than other kids and never needed medical intervention beyond simply taking some children's Tylenol. He had been in daycare since he was 9 months old, where they were engaging him in early learning. He should have picked up speaking but that wasn't quite happening. He could read from when he was two and a half and we thought he was just too playful and impatient to slow down and learn how to talk.

Around this time my wife quit her job to stay home and help the kids full time. It was tough to teach a kid that wouldn't sit or stand in one place. Finally, when he was 5, he was formally diagnosed by a special needs preschool psychologist as falling on the autism spectrum.

In the meantime, his younger brother was also starting to show signs of not developing language. He was not speaking full sentences and at 3 years old he was also identified as having speech development delays and admitted to special learning school. Life suddenly became very busy with 2 kids on the Autism spectrum. We were running between various doctor offices, specialists, speech therapists, ABA therapists, and trying out various classes including swimming, karate, piano, and gymnastics. They simply did not fit into any group classes due to focus issues and their inability to listen and follow instructions. We left no stone unturned trying to help these kids, but were clueless as to what had happened to these otherwise very cute toddlers. Doctors were quick to say that autism was not very well understood and could happen due to hundreds of reasons. I was adamant about solving the autism puzzle for my kids but it was almost like we were hitting a wall and nothing positive was happening.

Finally, a diagnosis breakthrough came when a neurologist we had consulted ordered an MRI scan for our second little boy. The results were a total revelation, causing me to remember long-forgotten incidents that happened at the delivery room years earlier that led to the current situation. For the first time in years I had conclusive evidence from the MRI of exactly what had gone wrong: our boys had suffered traumatic brain injury at birth.

Magnetic resonance imaging is a technique used to obtain images of the various interior organs of the body by utilizing the interaction of the body with magnetism and radio frequencies. It relies on the fact that the human body is composed of 65% water and the ionized hydrogen atoms (protons) in water and fat molecules are specifically targeted by this technique.

Putting the hydrogen atoms through a heavy magnetic field causes them to align their random atomic spin directions along with the magnetic field. A radio frequency is then used to perturb this spin alignment, which happens when the frequency is in resonance with the spin. When the radio frequency pulse is turned off, energy is given off by the hydrogen atoms as they return to their previous positions. This energy is measured by a coil to create accurate images of the interior organs. "Sliced" views can be obtained from any angle with the resolution being quite high: on the order of millimeters for magnetic field strengths of 1.5 tesla. It is typical to generate the 3 axial imaging planes with a single MRI scan. Trained radiologists reading these images can identify abnormalities including scarring, structural deficiencies and the like.

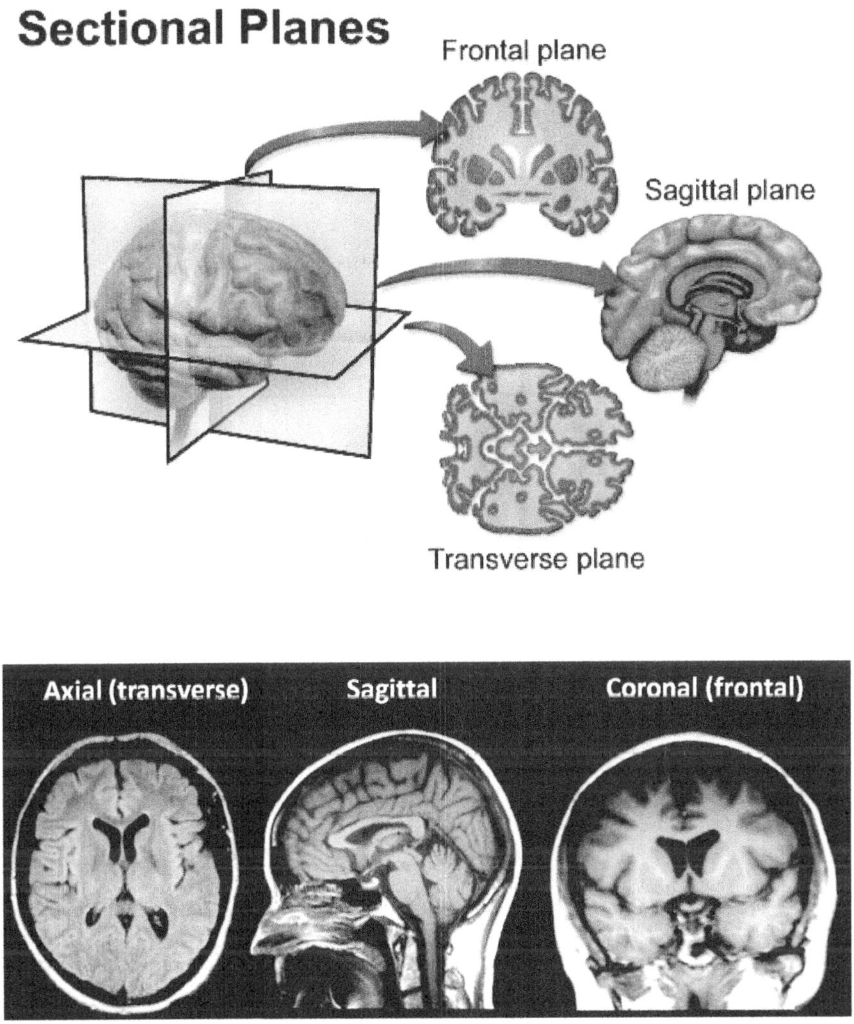

Fig. 5.1a, b: The three different planes in brain imaging with MRI.

Unfortunately, due to the young age of children and the need to be anesthetized for this procedure, most parents with children experiencing neurodevelopmental disorders have reservations about this procedure.

What was it about the MRI report of my younger son that triggered my recollection? See for yourself:

Children's Medical Center Dallas
1935 Medical District Drive, Dallas, TX 75235 (214) 456-2814

Radiology Report

Name: DOB: MRN:

Ordering MD: Location: Magnetic Resonance Imaging,

Exam Date/Time	Order Number	Accession Number
07/19/2016		

Procedure: MR BRAIN W/O CONTRAST

Reason for Exam: EXT-Delayed developmental milestones

Current Symptoms:

Report:

EXAM: MR BRAIN W/O CONTRAST

HISTORY: 7 years -old Male with EXT-Delayed developmental milestones Delayed milestone in childhood

TECHNIQUE: Sagittal T1 weighted images, axial FLAIR images, axial DWIs, coronal T2 weighted images were obtained.

Additional sequences: None

Field strength: 1.5T

COMPARISON: None

FINDINGS:

The corpus callosum is normally formed and of normal thickness , the 4th ventricle is normal in position, and there is no tonsillar herniation. There is a normal T1 hyperintense neurohypophysis within the sella turcica. The optic chiasm, tracts, and nerves are normal. There are no congenital midline anomalies.

The axial images show normal ventricular size and configuration. No pathologic extra cerebral fluid collections are present. There is a normal volume of white matter which is appropriately myelinated for age. There are scattered small areas of abnormal T2 prolongation within the peripheral frontal and parietal white matter, with a slightly larger focus measuring on the order of 1 cm in the left parietal white matter. These likely relate to multiple foci of white matter gliosis from remote insult. There is no evidence of cortical dysplasia, mesial temporal sclerosis, mass lesion, or hemorrhage. The posterior fossa is normal.

1 of 2

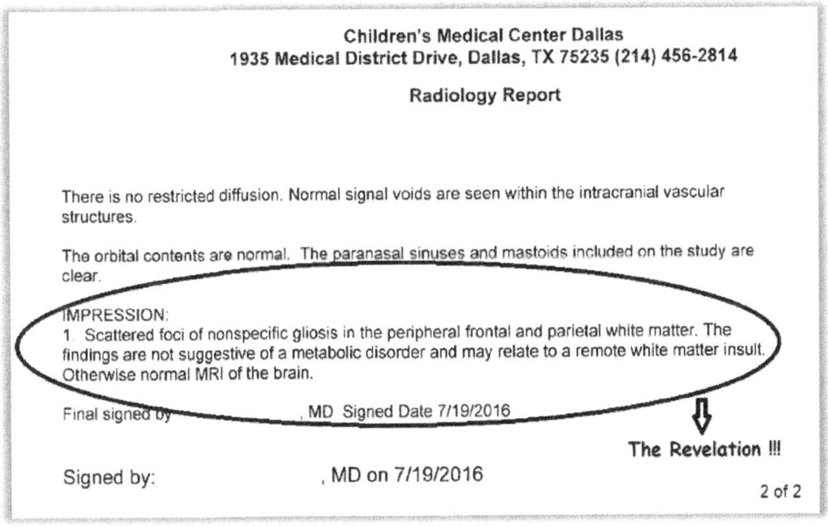

Children's Medical Center Dallas
1935 Medical District Drive, Dallas, TX 75235 (214) 456-2814

Radiology Report

There is no restricted diffusion. Normal signal voids are seen within the intracranial vascular structures.

The orbital contents are normal. The paranasal sinuses and mastoids included on the study are clear.

IMPRESSION:
1. Scattered foci of nonspecific gliosis in the peripheral frontal and parietal white matter. The findings are not suggestive of a metabolic disorder and may relate to a remote white matter insult. Otherwise normal MRI of the brain.

Final signed by , MD Signed Date 7/19/2016

The Revelation !!!

Signed by: , MD on 7/19/2016

2 of 2

Fig. 5.2: MRI report of my second son at age 7

As you can plainly read, this report states there were impacts to the brain of my second boy due to white matter damage – a phenomenon known as gliosis. Gliosis is the general term for "nonspecific reactive change" of glial cells in response to damage to the brain. According to Wikipedia, in most cases, gliosis involves the proliferation or hypertrophy of several different types of glial cells, including astrocytes, microglia, and oligodendrocytes. In its most extreme form, the proliferation associated with gliosis leads to the formation of a glial scar. In layman's terms, that means my boy suffered damage to the neurons in his brain, particularly to the axonal fibers which make up most of the white matter. As a reaction to such damage, the surrounding neuron support cells (called glial cells) have proliferated in certain areas, a sign of the body working hard to attempt damage control. Such traumatic damage to the deeply situated white matter of the brain is only possible during birth, when the head is subjected to high contraction forces – as in Pitocin induced labor. This is the "remote white matter insult" that the report is referring to.

Lost Memories Rekindled

Once I saw this MRI report, I started remembering things that had happened during the delivery of the boys. A feeling of guilt of having let this damage happen to our babies started taking up most of my waking hours and led to many sleepless nights. When my wife was induced for our first baby boy, about 5 to 6 hours into the process, the nurse had come into the delivery room carrying a phone handset, saying the doctor wanted to talk to us. On the other end of the line was the doctor who had demonized the C-section all along suddenly recommending that we consider one. She began by saying "At this point you can decide to proceed with the current course or have a C-section if you want," later adding that a "C-section is not that bad, let's do a C-section." She even had a giggle in her voice as she said it. She knew I was totally confused by her about-face as would have been expected going by her long-held view against C-section. I was pretty surprised by her sudden turnaround. I asked why and she simply said "in some cases it's not a bad choice." To this day, I wish she had given better reasoning, unfortunately, she did not offer anything further by way of explanation. Not convinced by her reasoning (or lack thereof), I responded that having waited this long could we continue for a couple of more hours and check back on progress. Her response was with a question "How is the mother doing? Is she tired? Can she go on for some more time?" My wife was doing perfectly fine as far as I could see and I told her so. Dr. S. then agreed to proceed with the current course. All in all, it was a 5-minute conversation that changed our lives and that of our child forever.

I was totally naïve to what was happening, but in reality, the nurse had warned the ignorant doctor about the risks of proceeding, but it ended up that the doctor had failed to explain her reasoning and ended up giving in to my request to continue for a couple of hours. The nurse couldn't believe the doctor gave in to my request. She simply couldn't hold back and as she was hooking my wife up to the next dose of Pitocin she made a prophetic remark: "Did you know this can cause brain damage?" At that moment I didn't even know if she meant brain damage to my wife or child, especially since the doctor had talked so much about the state of the mother a couple of minutes earlier. My wife looked fine to me and in my mind, I wrote off her comment as potentially some kind of adverse chemical reaction she was warning about. I told the nurse that the doctor didn't say anything like that, and dismissed the concern with a dismissive wave of my hand and a comment that would come back to haunt me: "Don't they write those warnings on every drug out there?" That was the end of the conversation.

I did tell her we had agreed to a couple of more hours, but the nurse and assistant nurse were upset at that time and I couldn't understand why. The older assistant nurse made a sad pout when she heard the decision. But this was pre- Affordable Care Act (Obamacare) times and nurses were not allowed to offer medical advice. I guess that was the reason they didn't pursue the matter any further. But the fresh attempt at induction, which I now know used a higher dose of Pitocin, brought on the dilation. A couple of hours later the nurse came back and sounded surprised when she reported that there had been progress. Eventually the baby was ready to be delivered after reaching 10-cm dilation around 8:30 pm in the night, a full 12 hours from the original start time, and he was born after an additional half hour of pushing.

Shortly after the baby was born, another incident happened. It was when the very experienced Pediatrician Dr. Joe examined the baby for the first time within an hour or two after he was born. Dr. Joe was visibly frustrated with what had happened. He was just done examining the baby at the when the nurse brought me over to him. He actually started yelling at the nurse that had brought me into the nursery room where he was examining our baby.

"I DON'T UNDERSTAND WHY THEY STILL KEEP DOING THIS, HE SHOULD HAVE BEEN C-SECTIONED OUT, WHO WAS THE OBSTETRICIAN?"

He then turned around to me and his exact words were

"His reflexes are poor, He should have been C-sectioned out, C-section is the best in these cases."

He continued with his reading to me on his assessments.

"The baby is fine, he is healthy overall, but his reflexes are weak, that's because he was induced for so long."

At this point I asked him, so when will his reflexes improve? Will he be ok in a few days? To that, Dr. Joe. did not answer but just gave me a helpless look and cast a helpless smile at my naiveté, and just shook his head and said, "I hope so, but it will take a long time." That was the end of our conversation. He didn't quite say how long that time was going to be and I didn't think much about it. Everything I had heard was that babies rapidly recover from most everything because they grow very, very rapidly.

The truth didn't register until I read the MRI results from my second son years later: my first baby had suffered brain damage as well, exactly as the nurse had warned. Dr. Joe's words and frustration now made perfect sense to me. The pediatrician was rightfully upset with the way events had turned out. He knew there were going to be issues down the line, which corroborates with what we are finding out years later. He had obviously not wanted to break the bad news that the baby had suffered permanent brain damage and ruin our happy 'new baby' moments. But because of what happened next, I wish I had deciphered the full extent of what he meant.

The Big Squeeze

After getting the MRI back on my second boy and the flood of revelations it triggered, I had an MRI performed on my first son. I wanted confirmation of what had happened so long ago to help me make sense of everything. There was just one hitch: when the result of his MRI scan was returned, it did <u>not</u> show any abnormalities in brain shape or size at 10 years of age. So, no brain damage from a Pitocin induced delivery?

Not so fast.

It turns out that there's a good explanation for why there's no visible brain damage on my first son's MRI. Recall that he was a big boy, weighing in at about 8.5 pounds, and had to be pushed out after the 10cm dilation of the cervix had been achieved by the 12 hours of Pitocin induction.

Now brain trauma would typically show up on an MRI for you or me if our brains were exposed to this trauma outside the womb. But for my sons – and for the thousands of babies that are born with autism – the trauma happens *inside* the womb as a result of Pitocin induced delivery. And there's one more thing that has to happen before a baby enters the world. It has to travel down the birth canal.

Now a baby's skull is a disjointed set of 5 bones designed for exactly this. It freely takes on a conical shape due to the massage of the even pressure from the mother's pelvic muscles in the birth canal. The brain is the consistency of butter at this stage, and so it is technically pretty fluid and is be able to take shape deformation without too much trouble at this early stage of life. And indeed, we would be poorly evolved creatures if the normal process of birth somehow caused brain damage.

Now unlike my second son, who as you'll see quickly popped out once dilation was achieved without any pushing whatsoever, there was just over half hour of pushing in the case of my first son before he was born. Unsurprisingly, due to the pushing his head was crowned and he came out with a conical elongation.

But when it comes to identifying the presence of damage, what has the most significant impact is the scar re-distribution of white matter damage. The effect of subsequent pushing means that any localized damaged tissue in the fluid brain is actually distributed over larger areas, thus erasing it from MRI scans (an MRI can only see any scarring or damage that is larger than 1 millimeter in size). In my older boy the pressure and deformation from the pushing had the effect of 'erasing' the scar tissue from being detected in an MRI scan by redistributing any damaged cells over larger areas of the brain.

Key Take Away: Diagnostic evidence in the form of MRI scan images fails to present itself as scarring in the brain in most cases because the pushing stage of labor tends to erase any scarring that has happened in the cervix dilation stage of labor by re-distributing the scar tissue such that it becomes invisible from MRI scan images. The first-born child and sometimes even subsequent children of a mother typically will not have scar tissue showing up in an MRI scan, if pushing is involved. This has allowed traumatic brain injury at birth to remain hidden for decades.

CHAPTER 6

OB/GYNs and Pitocin

Two years down the road we were pregnant with our second child. We went again to Dr. S.; after all, we weren't quite aware of any issues with our first kid, and for all we knew, Dr. S was an expert OB/GYN. But this time we already knew about induced labor. Dr. S. declared that since our first was induced, chances are the second would have to be too. And she was right. This time again, my wife was scheduled for induction on the due date. But unlike last time we didn't go back home; it was one long bout of 18 hours of Pitocin-based induction.

Delivery #2

Sometime during the induction process, I heard a couple of nurses talking outside the ward. "That Dr. S. is at it again; she has been inducing this poor couple for over 7 hours," said one nurse, with the other nurse replying "oh my goodness."

Moments later the head nurse suddenly rushed into the room and asked in a very declarative tone:

"DO YOU KNOW WHY SHE IS INDUCING YOU FOR SO LONG?"

We responded by saying we had our first boy who was induced born and so we were expecting this one to be the same. The nurse frowned and said "Is he OK?" to which I answered he was. The truth was our first son was hyperactive and not verbal at close to age 2, but we didn't think of it as an issue then. She replied with a "really?... hmm" and left with a frown and a confused look. This time there was no call from the doctor and no other interruption from the nurses.

I didn't quite know what to make of what the nurse was trying to say at that time. After 2 years I had completely forgotten the episode I had with the nurses and Dr. Joe with our first boy. Our induced labor continued on for a total 18 hours until our second baby boy came out. Being a smaller baby by a full pound, he did not have to be pushed but slid right out. He had a well-rounded head due to the lack of pushing. This lack of the pushing stage is also the reason his TBI and scarring is visible on his MRI scan.

His MRI scan shows scar tissue and foci of white matter damage distributed all over the brain where he was pressure damaged by the induction process. Due to the lack of pushing and the lack of pressure and crowning of the head, the damage to his brain remained intact, allowing it to show up as scar tissue in the MRI scan. The location of the scar tissue within the brain is also critical in the developmental delays. He has a long scar in the language area of the brain and directly correlates with his lack of ability to pick up language.

Pitocin Contra-indications

Brain damage at birth can happen even without induced labor, due to other reasons like hypoxia, infection, meconium, placenta conditions and so on. But when such complications do not exist and induced labor is pursued beyond the point of contra-indication, then it has a 100% certainty of brain damage. The reason I say this is 100% is because this is traumatic brain injury caused by the mechanical means of uterine contractions and not some kind of subjective chemical effect like in other drugs where the extent of an individual's adverse reaction to the drug might vary. Don't just take my word for it – that's exactly what it says could happen on the Pitocin label.

Reading through the warnings and contra-indications was an eye-opener and ended up giving me a clearer picture of what had happened to my kids and what has been happening across the US in general, causing the autism epidemic.

A contra-indication is defined as something (such as a symptom or condition) that makes a particular treatment or procedure inadvisable. The Pitocin package insert describes the following as contra-indications:

CONTRAINDICATIONS

Antepartum use of Pitocin is contraindicated in any of the following circumstances:

1. Where there is significant cephalopelvic disproportion;
2. In unfavorable fetal positions or presentations, such as transverse lies, which are undeliverable without conversion prior to delivery;
3. In obstetrical emergencies where the benefit-to-risk ratio for either the fetus or the mother favors surgical intervention;
4. In fetal distress where delivery is not imminent;
5. Where adequate uterine activity fails to achieve satisfactory progress;
6. Where the uterus is already hyperactive or hypertonic;
7. In cases where vaginal delivery is contraindicated, such as invasive cervical carcinoma, active herpes genitalis, total placenta previa, vasa previa, and cord presentation or prolapse of the cord;
8. In patients with hypersensitivity to the drug.

OB/GYN's not understanding or paying heed to this single line has been the cause of almost all autism in America and the world !!!

Fig. 6.1: Pitocin Label – Contra-indications section.

When you don't achieve any progress in dilation of the cervix in spite of 4-5 hours of continuous contractions at the optimal interval, further pursuit of induction as an option is contraindicated. Why? Because of the adverse reactions that could ensue. These too are plainly listed on the drug label:

The following adverse reactions have been reported in the fetus or neonate.	
Due to induced uterine motility: • Bradycardia. • Premature ventricular contractions and other arrhythmias. • Permanent CNS or brain damage. • Fetal death. • Neonatal seizures have been reported with the use of Pitocin.	Due to use of oxytocin in the mother: • Low Apgar scores at five minutes. • Neonatal jaundice. • Neonatal retinal hemorrhage.

Fig. 6.2: Pitocin Label – Adverse effects section.

Every one of the uterine motility adverse reactions is actually a manifestation of brain damage. Bradycardia is abnormally slow heartbeats which in turn leads to low oxygen supply to the brain causing hypoxia. The point about premature ventricular contractions also causes low blood supply from the heart. Permanent CNS or brain damage is what every fetus that has been subjected to prolonged induction has suffered, causing the autism epidemic. Seizures again happen due to traumatic brain damage and result in autism after birth. The Apgar score at birth is indicative of low reflexes and brain damage as was pointed out by Dr. Joe in our case.

While the warning on Pitocin is pretty clear and Pitocin contra-indication is laid out in the instructions for the drug, OB/GYN's have continued to prescribe Pitocin in contraindicated situations time and time again to this day. The American College of Obstetricians & Gynecologists (ACOG) should have come out strongly against this Pitocin abuse as strongly as they seem to condemn the abuse of C-sections, but instead seems to have remained silent. It's possible they are trying to protect their OB/GYNs from what they might see as unnecessary lawsuits for past deeds. While it is understandable that they might not want to scare people away from a treatment course that is noninvasive and beneficial to some, not highlighting the risks enough has resulted in malpractice that destroys the lives of victims, their families and impacts society at large.

Unfortunately, the reality seems to be more than just doctors protecting other doctors. Since the early 1980s the ACOG has been actively involved alongside the manufacturers of Pitocin in the development of the practice of inducing labor. The Pitocin label itself cites a study by the ACOG OB/GYNs, but the paper makes no mention of contra-indication, fetal brain damage possibilities and Pitocin warnings in that study or in their many publications (the closest they seem to come to Pitocin contra-indication is to talk about the adverse outcome of C-section if Pitocin induction fails to result in delivery). As a result, most obstetricians interpret adverse outcomes as fetal distress requiring C-section while ignoring the contra-indication time limit, resulting in the ever-increasing autism numbers.

The Blame Game

This book is based on the premise that in more than 90% of cases labeled as 'autism', 'ADD', 'ADHD', 'learning disabled,' etc., is really infant brain damage at birth. However, there can be multiple causes for brain damage unrelated to birth. But a closer look at these purported causes for autism reveals them to be poor explanations, leaving pregnancy and birth-related complications – which include a significant percentage of obstetrician malpractice involving Pitocin abuse.

For example, there are numerous chemicals in our environment that are toxic with the potential to cause brain damage. The predominant ones are mainly heavy metals including mercury, lead, arsenic, cadmium, chromium and the whole host of other chemical derivatives found in pesticides, building material, furniture, plastics, and so on. While their effects are true and they are undeniably dangerous, they have become the default scapegoats for everything that ails our modern society. They are now being used as the excuse for numerous health issues – including autism.

These chemical toxins can affect various parts of the body. They can be allergen, carcinogenic, cardiotoxic, genotoxin, hepatotoxic, nephrotoxic, neurotoxin, sensitizer, or teratogenic. From the autism standpoint we need to consider the effects of neurotoxins, which affect the brain and nervous system, and teratogenic agents, which cause birth defects.

With respect to teratogenic agents, birth defects tend to be severe, causing effects in many parts of the body, and typically more devastating than just cerebral shortcomings such as seen in autism. More broadly, the reason why damage from chemicals tends to be severe is because the human body is capable of repairing minor DNA damage; consequently, damage from the chemical needs to be really severe for birth defects to occur. For DNA damage to be passed on to a subsequent generation, damage needs to occur in the haploid gamete cells (eggs and sperm) which is unlikely to happen in the fortified, immune privileged organs like the testicles. Had this been the case, it would have been detected very easily; instead, after years of research science continues to come up empty-handed. It's important to note that the human body can tolerate trace amounts of various toxins pretty well and in many cases excrete them effectively. Any other exposure like the trace amounts someone may get from casual contact with furniture or building materials have to be accumulated without being excreted over long periods of time – possibly decades – for any negative impacts to show up. Even then the effects would show up later in the lives of individuals rather than the first few years. These also mostly end up causing cancers than neurologic defects.

People are right to be concerned about chemicals in the environment; but there are some telltale signs that are missing, in case chemicals were the cause of autism. For example, chemicals tend to show up first in smaller living animals with a shorter life span before its impact shows up in humans. No such major impact to the entire ecosystem across multiple species is being seen due to any of the chemicals currently being suggested as the cause for autism. Chemical contamination of the water supply, food sources or air also happens in clusters and the symptoms are universal and more obvious than just 1 or 2 in a 100 being affected by it. And if autism was caused by chemicals, people growing up in less industrialized countries with very little incidence of autism like myself would not be impacted. But I know numerous people who immigrated from practically every nation under the sun that have been affected because their kids were born in the United States. Autism is apparently neutral to parents' nationality.

Perhaps most damning to the chemical causation thesis is just how localized autism is. Newborn children are diagnosed within the first 2 to 3 years of their life with an affliction that only seems to impact their brain functioning. No impacts are found anywhere else in the body. But chemical effects (if congenital) would be more severe than involving just the brain. If it was chemicals other organ systems like the liver and kidney that are involved in the process of keeping the body free of chemicals would actually be the first ones impacted. However, no such effects are seen in kids with Autism, ADD and other attention deficiency and hyperactivity disorders. It's time to reject the chemical explanation of autism as not fitting the facts.

Another purported culprit is ultrasound scans. Some recent discoveries have led to casting doubt on the safety of ultrasound scans during pregnancy as a possible factor in fetal damage and autism. Ultrasound monitors for medical imaging were invented in the 1960s and have been used in fetal ultrasound scanning in the US since the mid-1970s. They work by the principle of the echo reflection of ultrasonic sound waves by objects. The ultrasound probe transmits sound pulses which then travel into the tissues of the body. Some of the pulse is reflected at tissue boundaries while some of it passes through, only to have other tissue boundaries reflect it. By accurately measuring the reflections and the time it takes to be reflected (to the millionths of a second), the ultrasound machine creates a depth and density image of tissues and organs in the body on the screen. Millions of pulses and echoes are sent and received each second.

It's important to note that ultrasound waves are not electromagnetic radiation and do not cause any ionizing effects on tissues at an atomic/molecular level. There are two ways the acoustic pulses might perturb the target tissue. They can bring about a temperature increase in the target area or can cause mechanical movement of the tissue in a phenomenon called acoustic cavitation (defined as the growth and collapse of pre-existing microbubbles in liquids under the influence of an ultrasonic field). Safety limits have been established in terms of a Temperature Index(TI) and a Mechanical Index(MI). Commercial ultrasound scanners are required to display these safety indexes on the screen at all times so the operators can stay under prescribed limits. And in fact the guidance given operators is to use the lowest possible probe output for the shortest possible time compatible with obtaining diagnostic information.

Experiments in pregnant mice have shown that ultrasound directed at the mice fetus during typical neuron migration times in development for prolonged durations results in irregular distribution of neurons in the brain after birth. However, the mice in these experiments were subjected to hours of ultrasound and such irregular brain development is easily identified in scans. Needless to say, human fetuses are not subjected to these levels of ultrasound, nor is this kind of irregular brain development characteristic of autism. We can reasonably conclude autism is not caused by the limited exposure to ultrasound that is the norm in human births.

The third and arguably most common claim for the cause of brain damage leading to autism is genetics. Yet every known genetic disorder afflicts more than just one organ in the body. For instance, Down's syndrome – one of the most familiar genetic disorders – creates visible facial features, heart conditions, developmental delays, higher cancer risk and a series of issues affecting all parts of the body, not just the brain and nervous system. Autism on the other hand only affects the higher-order functions of the outer lobes of the brain, the cerebral cortex and the multitude of functions controlled by these parts of the brain. No other organ is ever found deficient.

These genetic symptoms and diseases have also typically been described since historical times, while autism is a relatively new phenomenon, conflicting with it having a genetic cause. In the United States there exist communities of Amish people totaling about 250,000 who live their lives in the countryside in a very traditional way, away from the thrills and frills of modern living. In their rural communities, they grow their own foods and livestock, avoiding fertilizers, pesticides and other modern agriculture techniques. The Amish do not go to hospitals to have their babies and they don't vaccinate, and in these communities the incidence of autism is close to zero (as no scientific studies exist these estimates are based on people receiving services). Additionally, being a small closed community that descended from families of Swiss immigrants, they are susceptible to inbreeding and consequently have higher incidence of genetic diseases that are mostly unheard of in the general population – and there are many Amish receiving services for these genetic diseases. This pours cold water into the argument for the genetic causes of autism.

But perhaps most damning of all for the genetic argument is autism's frequency. Wikipedia lists over 230 different genetic disorders discovered so far that could impair brain development. Some of them as rare as one in 10 million or more. While such rare cases can be traced back to genetics, autism is far more common, occurring with every 1 in 60 births. Researchers trying to find genetic roots for autism always come up empty-handed and almost always point to the 'complexity' of the problem to explain their failure to find a genetic cause that essentially doesn't exist.

Finally, autism often gets blamed on bacteria or viruses. At a time when viruses like Zika have made headlines for causing mental retardation in unborn babies, it's not surprising to hear that viruses could be causing autism. But the point to note is viruses like Zika that impact the brain typically influence the growth of the fetal brain, resulting in a condition called microcephaly where there is a noticeable reduction in the size of the head due to reduced or stopped brain growth. Viral impacts to the brain are easily identified in an MRI scan as continuous lesions in the brain – again something not seen in the MRI scans of children with autism.

Historically we have seen many diseases that impact the brain like polio, smallpox, rubella, and scarlet fever. These have left numerous children paralyzed, deaf, blind, etc. Many of these diseases have been eradicated by vaccinations and so it's highly unlikely that some undetected disease lurks out there secretly causing autism. And as seen in most of these cases, the manifestation of the disease is much more severe than the purely speech and social development related impacts of autism.

What then is the cause of autism? Our experience and research point to a single cause. The nurses at the large Midwest hospital where our boys were born were aware that inducing beyond a certain point would result in brain injury. The hospital had developed its own protocol to avoid it. But our OB/GYN went against the hospital's established norms. The nurses tried to stop her in at least one instance and again questioned her strategy in the second, but unfortunately were unsuccessful. And as a result, our boys suffered TBI. That – and that alone – was the cause of their autism and they are not alone.

Key Take Away. In this chapter I have clearly outlined with proof how my 2 boys suffered brain damage as a result of Pitocin abuse. Use of Pitocin for inducing labor after the loss of amniotic fluid needs to be restricted to 4-5 hrs. and if no progress is achieved in dilation during that time a C-section should be performed. This is the direction under the FDA sanctioned Pitocin labeling. However, OB/GYNs are either unaware or disregard this warning and continue to induce, causing brain damage in babies at an ever-increasing rate to this day. My observation (based on limited data collected) is that between 35-40% of autism is attributable to identifiable birth complications, but OB/GYNs who are unaware of the warning label about Pitocin contra-indication or do not pay it heed are by my estimates responsible for close to 50% of autism seen in the U.S.

Part III

The Impact of Vaccination

Numerous parents of children impacted by autism have observed significant retrogression in the skills of their children after vaccinations. If observation is the basis of all science, the scientific community has not been able to provide a satisfactory explanation for this observation. Is it just a case of these observant parents being paranoid or are there other explanations?

CHAPTER 7

How Vaccines Work

I had almost completed this book when the Covid-19 pandemic suddenly turned life upside down all over the globe. Almost overnight as a society we suddenly went from one casting doubts on vaccines to one begging for a vaccine to counter a fast-spreading virus. It has disrupted life in every part of the world and brought on financial ruin and many unnecessary deaths. It has served as a reminder of how vaccines remain our most potent defense in the fight against viruses.

The knowledge level of common people about vaccines and medical research has also hit a new high, making us more aware of the complexities involved in developing a vaccine. Terms like pandemic, herd immunity, zoonotic, immune privilege, immune suppression, etc., that were not common knowledge prior to the pandemic are now more widely understood, making some of my explanations that follow much easier to understand and perhaps even superfluous.

But Covid-19 has also brought out varied reactions in people, mostly bringing out the best in people, but also highlighted the low tolerance some have for the slightest of inconveniences. It has emphasized the desperate need for people to have social interaction for maintaining mental health and sanity. It has also brought to the forefront the varied effects a viral disease can have on different segments of the population. We became aware of the inflammatory syndrome inflicted on children by the Covid-19 virus. We learned of how Covid-19 specifically wiped-out entire Alzheimer wards of senior living facilities while the rest of the facility escaped relatively unscathed.

Then there's also been the insanity of Covid-19 parties, proof that the deadly and permanently-debilitating effects of viruses from the past have been completely lost on the current generation. It's a tribute to the success of vaccines while the pandemic has revived memories of the deadly effects of viruses on human life and health.

I see all this as direct evidence of the arguments I am making in the subsequent pages. I have not changed any of my content as a result of the pandemic as it stands the test of these new times.

Vaccines and the Immune System

Vaccines have had a checkered past when it comes to safety, and have for a long time been the suspect in the search for the causative factor in autism. Numerous parents have reported massive regression in the development and behaviors of their kids after vaccine doses. There have been many cases where kids that were seemingly developing normally suddenly developed autism symptoms right after their vaccine dose. These occurrences have been the single largest reason why an increasing number of parents refuse to vaccinate their kids. In fact, that is the reason I will dedicate a whole chapter about my observations with vaccinations and their role in the entire autism puzzle. But before we go about looking at causes and effects, it important to know about the immune system and vaccines in general: what vaccines are, what they are made of, and the basic mechanisms used to create these vaccines.

Before looking at vaccines let's review our very simplified representation of the immune system and about how immunity works previously discussed in chapter 3.

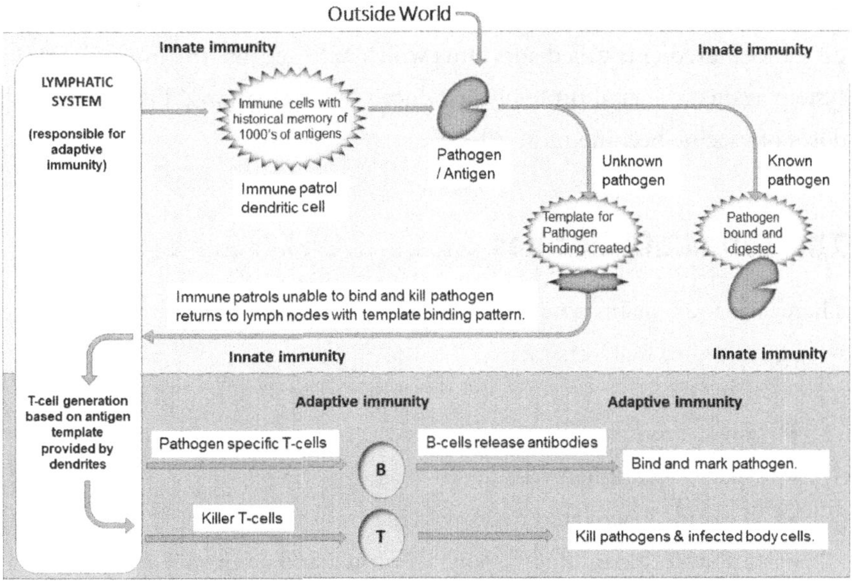

Fig. 7.1: Simplified schematic of the immune system

There's the first line of defense from invading microbes called innate immunity which comes from some of the basic barriers in the human body like skin, fluid in the eye, saliva, etc. These barriers, working alongside the dendritic immune cells at the tissue level with historical evolutionary knowledge capable of distinguishing body cells from invading microbes, provide a certain level of protection called innate immunity.

The second line of defense is the adaptive immune system involving the lymphatic system in the body. This adaptive part of the immune system provides the capability to fight new kinds of invasive microbes that the body has no previous knowledge of.

The concept that there is immunological memory in our bodies to disease-causing viruses and bacteria that can last anywhere from a few days to a few years to a lifetime is what is used to our advantage in designing vaccines. Vaccines are designed to create an immune response in the adaptive part of the immune system. This whole immune response works best when the adaptive immune system is activated by the innate immune system. Modern vaccines typically use small amounts of smaller parts of the virus or viral protein which may not elicit an immune response. Hence vaccines sometimes use other ingredients called adjuvants which help activate the innate immune system against the viral proteins introduced by vaccinations. This helps small doses of vaccine become more effective.

The Types of Vaccines

There are four main types of vaccines based on their composition and strategies used to make them:[1]

- Live, attenuated vaccine
- Inactivated/killed vaccine
- Toxoid (inactivated toxin) vaccine
- Subunit, recombinant, polysaccharide, and conjugate vaccines

Table 7.1 lists the popular vaccines and the category they fall under.

Vaccine Type	List of Diseases
Live, attenuated vaccine:	Vaccinia (smallpox) Measles, mumps, rubella (MMR combined vaccine) Varicella (chickenpox) Influenza (nasal spray) Rotavirus Zoster (shingles) Yellow fever
Inactivated/killed vaccine:	Polio (IPV) Hepatitis A Rabies
Toxoid (inactivated toxin) vaccine:	Diphtheria, Tetanus (part of DTaP combined immunization)

Subunit/conjugate vaccine:	Hepatitis B Influenza (injection) Hemophilus influenzae type b (Hib) Pertussis (part of DTaP combined immunization) Pneumococcal Meningococcal Human papillomavirus (HPV)

Each type has strengths and limitations in its approach to inducing immune memory against specific types of viruses and bacteria.

Viruses cause disease by quickly multiplying to thousands of units per ml inside the body of a natural host. To create weakened viruses for human vaccines, one strategy is to create an active live human virus population and introduce it into a new host or medium like eggs, cultured tissue, or even other animals. Viruses are typically very selective to the host, and a human virus cannot thrive in one of these mediums. However, there tends to be some mutation in some of the original population that allow them to survive and multiply very slowly in this new medium. This creates a slightly modified strain of the original virus, one that is a slow multiplier. The new strain – when re-introduced in the original host – only retains this ability to multiply very slowly and does not have the ability to cause disease. This new strain of vaccines virus is what live vaccines are made up of. However, these so-called weakened vaccine viruses replicate well enough in humans to induce "memory B-cells" that protect against infection of the actual disease virus in the future. The advantage of live "weakened" vaccines is that one or two doses provide immunity that is usually lifelong. The limitation of this approach is that these vaccines usually cannot be given to people with weakened immune systems.

Killed virus particle vaccines take an entirely different tack. Using this strategy, viruses are completely inactivated (or killed) with a chemical. By killing the virus, it cannot possibly reproduce itself or cause disease, but because the virus is still "seen" by the body, cells of the immune system that protect against disease are generated. Advantages include the fact that the vaccine cannot cause even a mild form of the disease that it prevents and that the vaccine can be given to people with weakened immune systems. However, the limitation of this approach is that it typically requires several doses to achieve immunity. There are also concerns about the quality control and safeguards on the kill process used in manufacturing, where if it's not a hundred percent thorough it could result in the live virus being administered as a vaccine.

A third approach uses bacteria. Some bacteria cause disease by making a harmful protein called a toxin. The toxin invades the bloodstream and is largely responsible for the symptoms of the disease. The protein-based toxin is rendered harmless (the toxin, once inactivated, is called a toxoid) and used as the antigen in the vaccine to elicit immunity. To increase the immune response, the toxoid is added to aluminum or calcium salts, which serve as adjuvants.

Subunit recombinant, polysaccharide, and conjugate vaccines – the fourth type of vaccines – is the classification type for many of the more recently developed vaccines, including DNA and RNA based vaccines. They use only a part of the virus in their design. The vaccine is composed of a protein that resides on the surface of the virus. This strategy can be used when an immune response to one part of the virus (or bacteria) is responsible for protection against disease. In bacterial vaccines it is also possible to use part of the sugar coating (or polysaccharide) of the bacteria. Protection against infection by certain bacteria is based on immunity to this sugar coating (and not the whole bacteria). However, because young children don't make a very good immune response to the sugar coating alone, the coating is linked to a harmless protein (thus a "conjugated polysaccharide" vaccine). In fact, many viral vaccines are also conjugate vaccines where a weak antigen is combined with a strong antigen so the immune response to the weak antigen in the vaccine is strong. One advantage of these subunit vaccines is that they can be given to people with weakened immunity as they don't contain the whole virus and therefore are not capable of causing the disease they vaccinate against. But the limitation of these vaccines is that they often need multiple doses to provide immunity and may require additional booster shots to get ongoing protection.

The reason I am offering this explanation about the making of vaccines is because the argument about the safety of vaccines is an emotionally charged topic. Numerous parents have noticed retrogression of previously mastered skills in their children after vaccinations, and feel strongly about them being the cause. Scientists and authorities keep brushing aside these concerns as coincidental or come up with other feeble excuses. If observation is the basis of all science, then this observation by parents needs to be acknowledged and explained, which has not happened so far. So, the question of vaccination safety for newborn kids remains a hard one for many parents.

Parents who know more about their child's health condition from birth may be in a better position to make informed decisions for their children if they have a better understanding of vaccines. Knowledge of the design and make up of these vaccines equips parents with important details needed for making just such an informed decision. It turns out that opposition to vaccines is typically directed against live virus vaccines and not against some of the others. Understanding why and how that manifests itself in the current controversy requires learning something of the history of vaccines.

Key Take Away: Covid-19's effect on Alzheimer patients reminds us of the devastating effects viruses can have on people with pre-existing brain conditions. The immune response to disease is comprised of innate immunity and adaptive immunity. Vaccines are designed to trigger immunological memory by activating the adaptive part of the immune system. We have seen how vaccines are classified into multiple types depending on how they are synthesized and the different strategies they use to create immune memory.

CHAPTER 8

The History of Vaccines

Vaccines have come a long way from the time when the first modern vaccines started making their appearance in the nineteenth century. Before then there were practices that amounted to vaccination against smallpox in China as early as 1000 AD and other such ancient practices were also known in Turkey and Africa.

The Enormous Success of Vaccines

The English physician and scientist Edward Jenner is credited with discovering the first modern vaccine in 1796 when he used cowpox (a disease similar to smallpox that infects cows) to create the first smallpox vaccine. Smallpox was known to kill up to 35% of people it infected and leave others scarred and blind. Over the next 200 years developments in vaccines and adoption around the world has resulted in smallpox being effectively declared eradicated from the world.

The practice of developing vaccinations for diseases accelerated after 1885 when French Chemist Luis Pasteur developed the first rabies vaccine. Soon vaccines were developed for numerous debilitating and deadly human diseases including vaccines for typhoid (1899), cholera (1911), diphtheria (1914), tuberculosis (1921), and tetanus (1924). In the mid-twentieth century vaccines for polio (1955), measles (1963), mumps (1967), and rubella (1969) were developed.

The polio vaccine was a classic example where the horrible disease led to its very swift adoption. In 1951 Dr. Jonas Salk discovered a way to create the polio vaccine; in US trials in April 1955, it was shown to be 80-90% effective. The vaccine was approved same day and in production by May 1955. Production methods were adjusted after a brief hiccup where some getting the vaccine were paralyzed, and the vaccine went on to reduce the number of paralytic polio cases to negligible levels within a decade. By 1994 polio was declared eradicated in the Americas.

With such a track record of success, vaccines became the most powerful disease fighting option in the hands of health officials. But typical immunity provided by many vaccines is not 100%. In many vaccines, effectiveness is more like 85-90%, so if an outbreak occurs it's still possible for some vaccinated individuals to be susceptible to disease, which is one reason health officials try to get as many people immunized as possible so as to keep any spread of disease under check. For obvious reasons it is also only possible to eradicate viruses that are exclusive to humans. In cases where the virus crosses over to other species or when the disease agent hides successfully in the environment (like in the case of food or water-borne diseases), it may not be possible to completely eradicate it. Some viruses can be particularly insidious and change frequently. This is the reason why we still need vaccines against some of these dangerous diseases even though many of the worst ones have been successfully eradicated. Vaccination is truly a glaring example of an instance where modern medicine has successfully triumphed over diseases that have plagued human beings for centuries.

While vaccine science has been advancing at a very rapid pace in recent times, our understanding of immunity and vaccines is still vague at best. Vaccines, particularly those involving live weakened viruses, are known to be risky when administered to immune-compromised individuals. On rare occasions they have been known to cause significant adverse reactions even in seemingly healthy individuals. Interestingly, the most significant side effects involve unwarranted immune reactions – immune attacks on specific tissues and cell types in the body. The most dangerous reactions are those that involve the nervous system and heart muscles, including anaphylactic shock. While any reactions might subside over time, certain effects can linger for a lifetime.

The Anti-Vaccine Movement

Opposition to vaccinations has existed for almost as long as vaccinations have existed. In fact, opposition predated vaccine injections during the times when the practice of variolation was the norm. But reasons for opposition have changed and evolved over time. Initially it was against the adverse effects of contaminated vaccines that ended up taking lives. Then there was opposition to the use of animal sources, mixed in with religious beliefs about interfering with God's agenda and fate. Add to this alternate scientific beliefs and political agendas and there were plenty of reasons to go around as to why someone might be opposed to vaccines. But the net effect of these movements was that it made vaccines better and helped improve and streamline the vaccine manufacturing and administration process in general.

Fast forward to modern times and you find that most opposition to vaccines comes as a result of their implication in causing modern afflictions like allergies, auto immune disorders and autism. Since some of the killer diseases have been eradicated, it might be expected that the number of necessary vaccinations might drop. Instead, more and more diseases have been added to the list, including non-eradicable ones like the flu. These are perceived by some as attempts by the pharma industry to profit from a continuous revenue stream by introducing new and unnecessary vaccines. There are thousands, perhaps millions of viruses out there, and some of them can cause communicable diseases in humans while others keep mutating and crossing over from animals as well. How many diseases to vaccinate against and which ones to assume risk on, are hard to define and an ongoing source of debate.

The profile of the modern parent who has concerns about vaccinations is contrary to what pro-vaccine folks would like to portray them as. Parents who don't vaccinate kids tend to be affluent, better educated and typically people who want to do the right thing for their kids. These people understand the difference between live and other forms of vaccines and in the absence of evidence or explanation for adverse outcomes do not want their child to be a guinea pig when outcomes are almost luck of the draw. Their beliefs are reinforced by parents that notice retrogression of skills in their children after vaccines. As a result, clarifying the circumstances around vaccine injury and helping determine better criteria for vaccine contra-indication is a pressing issue.

Where to Go From Here

There's obviously been a lot of sweat and blood invested into the development of vaccines over the past 200 years and it hasn't come easy. Researchers and health officials who remain exceptionally aware of these struggles and hard-fought gains in the history of vaccine development continue to be the most ardent supporters of vaccines. Achieving a world free of deadly and debilitating diseases like smallpox and polio has been no mean feat, and pro-vaccine supporters fear these victories are being forgotten when the vaccination rates in the population start falling. But what's lost in the pro-vaccine lobby is the understanding that historically, the challenges put forward against vaccines have actually led to improvements in safety and other positive changes in their manufacture, distribution and use. Researchers are now being challenged with achieving what will truly be the next big chapter in the history of vaccines in developing better alternatives to live vaccines (and ridding vaccines of at least some of the allegations that have been lodged against them). In the age of artificial intelligence and genetic engineering this sort of development remains in the realm of possibility, given the right motivation to make it a reality.

At the same time, government public health officials often worry about contagious diseases reaching pandemic proportions – a situation they are tasked with avoiding. In this day and age, diseases can pretty much travel at the speed of airplanes. As Covid-19 has illustrated, controlling an outbreak of disease by isolating it at the source, particularly the very infectious and deadly viruses, can be a very difficult challenge for health officials anywhere in the world. Their options are pretty limited, with the main tools in their arsenal either to quickly figure out a cure or get a vaccine ready that can prevent the spread. There are limits to how much physical methods such as quarantine and controlling other modes of transmission can help.

Vaccines therefore have become a vital part of any government's repertoire of tools to prevent disease pandemics. They have been widely adopted around the world as the primary defense against what are now called 'preventable' diseases. Deadly and debilitating diseases like smallpox and polio have been eradicated thanks to the efficacy of vaccine programs around the world. The cost benefits and incidental benefits have been so real that it's hard to argue against the value they have brought on to modern society. The upside to vaccines are so many that the occasional issues with their use, particularly things like allergies and other side effects, have been relegated to the fringes. Governments simply cannot afford to have people unvaccinated just because of seemingly remote possibility of side effects. They tend to fight really hard against any notion that vaccines are to blame for any injury caused to an individual.

It is widely agreed that vaccine production needs strict quality controls and any lapse could lead to actually giving the disease to the vaccinated (at least in the case of live vaccines). Many such instances have happened throughout the history of vaccines, like when the polio vaccine caused paralysis in the early days of its adoption. In addition, dangerous concoctions have been produced and distributed as vaccines by unscrupulous charlatans in the past, which resulted in deaths and other tragic outcomes. As a result, vaccine production has been brought under the control of government health agencies over time. This has mostly alleviated any risks associated with vaccine production and quality control.

But like any medicine, vaccines come with a lot of warnings noted on their labels and issues can crop up from time to time. In spite of their vastly improved safety, the warning label states that in certain rare occasions they do cause injury. But when impacted parents started taking on the vaccine manufacturers in court, manufacturers started leaving the business, in turn threatening the vaccine supply of the US. This prompted the US government to intervene with the National Childhood Vaccine Injury Act of 1986. Subsequently the Vaccine Injury Compensation Program (VICP) was created in the U.S. as a no fault alternative to the court system in an effort to bolster the vaccine supply while still providing out of court settlements for the most seriously impacted. The program investigates vaccine adverse effect complaints and approximately one in a million vaccinated cases have been receiving compensation.

In the late 1990s and early 2000s, more and more parents became aware of the program and worries increased about the perceived link between autism and vaccines. This resulted in more claims and increased success of cases regarding autism in vaccine court. Per the hrsa.gov website, in the 29 years from the time of creation in 1988, compensation to the tune of 3.7 billion has been paid.

As I will explain in the next chapter, visible retrogression happens after vaccines are administered in cases where there is prior brain injury. Although they are not the root cause, vaccines were blamed for autism in the early to mid-2000's and the rise in compensation demands threatened to destroy the VICP program. Over five thousand cases had been filed and were pending in US courts against vaccines prior to 2009.

To deal with the influx the government grouped these cases together into 6 representative class action suits and instituted a special court called the "Vaccine Omnibus hearings" to deal with them. The primary complaints were that mercury containing adjuvant (thimerosal) in vaccines and the MMR vaccine were causing autism. The 6 representative cases were fell into 3 biological categories: MMR only, thimerosal only and combination of MMR-thimerosal as the cause in autism. The court denied a "preponderance of evidence" in all cases and absolved vaccines of any responsibility for autism. The loss of the plaintiffs in these Vaccine Omnibus hearings in 2010-2011 dealt a severe blow to the anti-vaxxers whose real-world observations did not pass the legal test. But the court did not stop there: they pointed at genetics as the cause for autism. In our world, it seems easiest to blame unknown causes as "genetics" rather than accept the truth as "unknown." This move by the special vaccine court ended up diverting valuable research dollars chasing genetic causes that do not exist.

Key Take Away: Vaccines have been enormously successful in reducing and eliminating the occurrence of a whole host of debilitating diseases in the 20th century. This great success has encouraged the pharma companies, medical community and health officials to push more and more vaccinations on the population, unmindful of the adverse effects they have on certain people. The anti-vaccine movement in modern times has been a response to the adverse effects people believe they suffer as a result of vaccinations, such as allergies and auto immune disorders. The failure of the autism omnibus proceedings in proving vaccines as the cause of autism has further strengthened the hands of the pro-vaccine crowd, while the anti-vaxxers continue to be wary of them due to their observations of retrogression in skills of their kids or occurrence of severe allergies and immune disorders after vaccines.

CHAPTER 9

The Vaccine Double Whammy

As much as vaccines have been an enormous success, parents have in some instances seen adverse reactions in their kids after vaccinations. Their views often get written off as nonsense by the pro-vaccine advocates in the media and are dealt with tougher mandates from health officials and the government, making it difficult to avoid vaccinations. After all, vaccine safety has been tested time and time again and adverse reactions are mild in the vast majority of cases while more severe reactions can be attributed to other causes. But it's important to note that the anti-vaxxers are not all irrational idiots like the advocates and health officials and ill-informed media would like to paint them as, but rational everyday folks that were observant enough to notice cause and effect in their kids related to vaccinations. These are mostly people directly impacted by autism and developmental challenges in their kids or their immediate family and friends. The mental effects of autism emerge over a long time frame as they relate to childhood development and sometimes vaccinations are the trigger for realizing what is really happening.

That was certainly the case with me. My kids were not very verbal when they were vaccinated prior to age 2, and by the time I realized that the 3rd year vaccinations had sent them back much further it had been a few months. The only adverse reactions I immediately noticed were skin rashes in my second son (which we were accustomed to since it had happened every time he received a vaccination). The pediatrician's office prescribed topical creams to take care of it and to our thinking that was the end of the story.

I was unaware of the adverse history of vaccinations in the US and the presence of the VICP and VAERS adverse effect reporting system prior to writing this book. In fact, most parents I have spoken to do not know of the existence of the VICP or the VAERS programs. Even if I had known about them chances are, I wouldn't have reported anything – it was just a skin rash after all. Parents have to know about autism, anticipate and watch for it carefully at such an early age to be able to notice the effects. Health officials and vaccine advocates often point to the lack of reporting in the VAERS as a factor in their justification of the lack of adverse reactions. But even for the most observant parents the lack of awareness of the existence of these systems and the difficulty in anticipating the invisible mental and psychological effects on kids 3 years old or less is daunting. The cause and effects are hard to identify and correlate; unless you believe autism is caused by vaccines and watch for it, it's hard to catch. If and when retrogression happens by the time you do realize what's happening it's often much later.

The MMR Vaccine

Most vaccines do come with warnings around the potential for injury, but medical practitioners are mostly trained to push vaccines on everyone with their 'vaccines are safe for everyone' mentality and tend to not take any of these warnings into consideration. These vaccines are typically administered by nurses and not doctors, and even if nurses did want to warn patients, they are not allowed to offer medical advice.

Yet the proof of the danger live vaccines like the MMR represent for brain damaged individuals is pretty much hidden in plain sight. The first line of the MMR warnings section clearly states prior brain injury being a risk:

WARNINGS

Due caution should be employed in administration of M-M-R II to persons with a history of cerebral injury, individual or family histories of convulsions, or any other condition in which stress due to fever should be avoided. The physician should be alert to the temperature elevation which may occur following vaccination (see ADVERSE REACTIONS).

Hypersensitivity to Eggs

Live measles vaccine and live mumps vaccine are produced in chick embryo cell culture. Persons with a history of anaphylactic, anaphylactoid, or other immediate reactions (e.g., hives, swelling of the mouth and throat, difficulty breathing, hypotension, or shock) subsequent to egg ingestion may be at an enhanced risk of immediate-type hypersensitivity reactions after receiving vaccines containing traces of chick embryo antigen. The potential risk to benefit ratio should be carefully evaluated before considering vaccination in such cases. Such individuals may be vaccinated with extreme caution, having adequate treatment on hand should a reaction occur (see PRECAUTIONS).{46}

However, the AAP has stated, "Most children with a history of anaphylactic reactions to eggs have no untoward reactions to measles or MMR vaccine. Persons are not at increased risk if they have egg allergies that are not anaphylactic, and they should be vaccinated in the usual manner. In addition, skin testing of egg-allergic children with vaccine has not been predictive of which children will have an immediate hypersensitivity reaction...Persons with allergies to chickens or chicken feathers are not at increased risk of reaction to the vaccine."{47}

Hypersensitivity to Neomycin

The AAP states, "Persons who have experienced anaphylactic reactions to topically or systemically administered neomycin should not receive measles vaccine. Most often, however, neomycin allergy manifests as a contact dermatitis, which is a delayed-type (cell-mediated) immune response rather than anaphylaxis. In such persons, an adverse reaction to neomycin in the vaccine would be an erythematous, pruritic nodule or papule, 48 to 96 hours after vaccination. A history of contact dermatitis to neomycin is not a contraindication to receiving measles vaccine."{47}

Thrombocytopenia

Individuals with current thrombocytopenia may develop more severe thrombocytopenia following vaccination. In addition, individuals who experienced thrombocytopenia with the first dose of M-M-R II (or its component vaccines) may develop thrombocytopenia with repeat doses. Serologic status may be evaluated to determine whether or not additional doses of vaccine are needed. The potential risk to benefit ratio should be carefully evaluated before considering vaccination in such cases (see ADVERSE REACTIONS).

"Due caution" out here is a very vague term. This should mean avoid MMRII for individuals with with brain injury altogether

Fig. 9.1: MMR combined vaccine label

Admittedly the phrase "due caution" in this statement is still pretty vague and only points to stress due to fever and not to direct impacts of live vaccine viruses on injured brains. Phrases like that lend themselves to faulty interpretation by pediatricians brainwashed into believing vaccines are fine for everybody and people who avoid it are only being irrational. The label should clearly state that the vaccine needs to be avoided for pre-existing brain injury.

Of all the vaccines children are given these days, the measles vaccine holds a special place in the minds of anti-vaxxers. This is due to the observations of immediate retrogressive effects on their children by autism and ADHD impacted parents. My research into the literature helps me come up with the reasons why this is the case.

Live Vaccines & Viral Latency - Viruses that come and never leave

The measles virus, like the polio virus and certain others, is capable of attacking the endothelial cells that form the blood brain barrier directly and get past the tight junctions of the endothelial cells of the barrier. As a result, the measles virus establishes a lifelong latency in the neuron cells of the body. Viral latency is known to inhibit/consume the nerve growth factor (NGF) at least in the case of Herpes Simplex[9.1] (HSV1) and could be similar with measles, which at the early ages potentially stops brain development in its tracks and turns the clock back on development. Such a lifelong latency could result in enduring effects for the life of the impacted individual.

Viral latency can even occur in the neuron cells of the brain, when the live virus in the live vaccine interacts with the neuron cells of the brain, by getting past a damaged blood brain barrier. Such latency is probably causing what is sometimes referred to as mitochondrial dysfunction. Mitochondria are organelles in the cells that are responsible for generating cellular energy molecules called ATP. There are other very important functions including cellular signaling functions that mitochondria are involved in. When viral latencies resulting from the compromised blood brain barriers of a damaged brain interact with these various functions in neuron cells, they cause whole body dysfunctions at various levels, including brain fog, lethargy/hyperactivity, and general neuro muscular weakness in individuals suffering from autism from traumatic brain injury at birth.

Some of the viruses that infect the human body (particularly the herpes virus family but also many other virus types) establish a lifelong latency and never get eradicated from the body. This leaves open the possibility that they become active later in life or just sit around in certain body cells causing other diseases where the causative factors haven't been established. They do this by putting out material in cell cytoplasm (episomal latency) or by injecting material in the DNA in the cell nucleus (proviral latency). Certain viruses of the herpes virus family (herpesviridae) including chicken pox, Herpes simplex viruses (HSV-1, HSV-2), Human herpes virus 6 (HHV-6A, HHV-6B) all establish episomal latencies in neuron cells and are suspected in a lot of diseases where the disease associations haven't been proven. Proviral latency is the reason the HIV-AIDS virus has remained almost impossible to completely cure.

Interestingly, it's not only the live vaccines that come with this warning, but even other vaccines like DTP come with a warning of neurological side effects. In these cases, it's potentially the immune response, fever and predisposition to seizures that seem to be a factor. This is where the risk of high fevers that can literally 'fry the brain' causing damage and functioning as the cause of neurological issues. However, in these cases, the effects may not bring on lifelong debilitations. Only live viruses capable of surviving in immune privileged areas affecting neurons and establishing a neuronal latency are suspects in stopping of the brain healing processes following exposure to them.

Needless to say, live vaccines as an immunization strategy definitely need to be re-considered. These live vaccines and their production methods have remained the same for decades and in the wake of newer developments in the areas of protein and genetic engineering it's time to reconsider their use. But until a way around live vaccines is found; they are probably still something that will stay with us as the preferred method of health officials. It's true that alternate methods do not afford the lifelong immunity that live vaccines provide, but lifelong immunity also means introduction of viruses that establish a latency in the body of the immunized.

Autism by live Vaccine Virus & TBI Interaction: A Hypothesis

Dealing with 2 kids on the spectrum and by virtue of my first-hand observation of their development right from pregnancy and birth – combined with my research into anything and everything out there that could help identify the roots of my kids' problems – have led me to arrive at some conclusions on the exact etiology of their condition. My kids sustained traumatic brain injury by way of Pitocin abuse. Even though they were already delayed as a result of their birth injuries, their damaged brains tried to recover and they were beginning to show some progress with language and interaction. The vaccinations they were subjected actually kept sending them backwards at every stage with the eventual loss of those minimal verbal interactions after the third-year vaccinations. Looking back, they were improving and showing signs of progress but then retrogressed with every bout of vaccinations until we stopped vaccinations when my second boy was age 3. To this day we still sometimes notice some retrogression of skills and increase in behaviors after bouts of simple viral illnesses like common cold. Now of course it is obvious that the interaction of viruses with their brains is having an impact.

We had some viral antibody testing done on our boys where we found my first son had a high level of measles antibodies while my second boy had a high level of HHV6 (Human Herpes Virus 6) antibodies. While measles antibodies might be from the vaccine viruses, the HHV6 is a virus that pervades about 98-100% of all human populations without causing adverse reactions. We now keep them on long term antiviral medication which helps them with their stimming behaviors and helps them focus better. Such antiviral medications are not a cure and do not result in the reversal of the damage caused by vaccines and autism; indeed they offer only a slight improvement compared to the significant retrogressions suffered after vaccinations.

Having talked to numerous parents about their experiences leads me to the conclusion that vaccines are the second part of a double whammy on a child's brain in the early years of life reinforcing the brain injuries that are the cause of autism. Kids with brains damaged during the birth process begin to recover (with girls recovering better than boys due to the brain protective nature of estrogen and potential redundancy in brain functionality in girls). But vaccinations stop that recovery in its tracks (particularly the live vaccines which alter the brain protein chemistry in unfavorable ways).

A damaged brain with large areas of injury sustained at birth is a maze of dead and damaged neurons, scar tissue, reduced blood flows, and a compromised blood brain barrier, lodged inside a body working hard to grow new pathways to heal the damage. The glial cells of the brain are responsible for both brain immunity and the local wound healing process. Glial scars are required to seal the site of injury, protect the damaged neural tissue, prevent an overwhelming inflammatory response, and to re-establish the blood brain barrier. But these same scars are known to inhibit the regrowth of neurons through the damaged areas.

The compromised blood brain barrier and scar tissue might represent entry points where viruses might penetrate the CNS including immune privileged areas like the white matter. Live vaccines and other ubiquitous worldly viruses potentially become firmly entrenched in immune privileged areas, scar tissue and other low blood supply areas of the damaged brain and successfully hide away from the reach of the body's antibody defense mechanisms. In this safe environment of the damaged brain, these viruses might also keep multiplying at a slow rate and keep spilling virus particles into the active parts of the brain and body, triggering constant antibody responses. It is possible that neurons surrounded by the constant presence of either virus or virus antibodies are deprived of nutrients and inputs required for their proper functioning. It is also possible the virus or antibody particles interfere with the synapses connecting neurons by accumulating at these electrically rich junctions, thus impeding electrical impulse signals from traveling through critical connection routes in the already damaged brain, resulting in the worsening of brain functionality with viral exposure. Add to this the ACAID phenomenon described earlier about how viruses can obtain whole body immune privilege by entering certain vital areas of the body and you can see why viruses in any form are potentially catastrophic for children with autism from traumatic brain injury.

There could be further complexities in how these mechanisms actually work. For example, it's known that viruses like measles already have a brain impact listed as one of their effects on the body. The statistics say 33 out of every 100,000 individuals who actually got measles end up with CNS damage. Diseases like measles, chicken pox etc., are known to permanently change your body by establishing a latency, causing long term effects like allergies in a person that previously did not have such allergies. The measles vaccine actually contains 13 different known variants of the measles causing virus, which might cause this allergy effect to be magnified even further. These allergies (particularly the food allergies) in turn obstruct the absorption cycles of vital nutrients by the body, resulting in an imbalance of chemicals in the body and brain. Such interference in absorption functions by these gut allergies at a critical time in brain development could then be another mechanism that can further negatively impact the body's attempts at recovery from brain injury. The relationships between TBI and gut dysbiosis have now been borne out by numerous recent studies, even in adults[9.4, 9.5]. Fig 9.2. summarizes this interaction between viruses possibly including the weakened live viruses found in live vaccines and the damaged brain and a pathway for vaccines and viruses in general to compound autism.

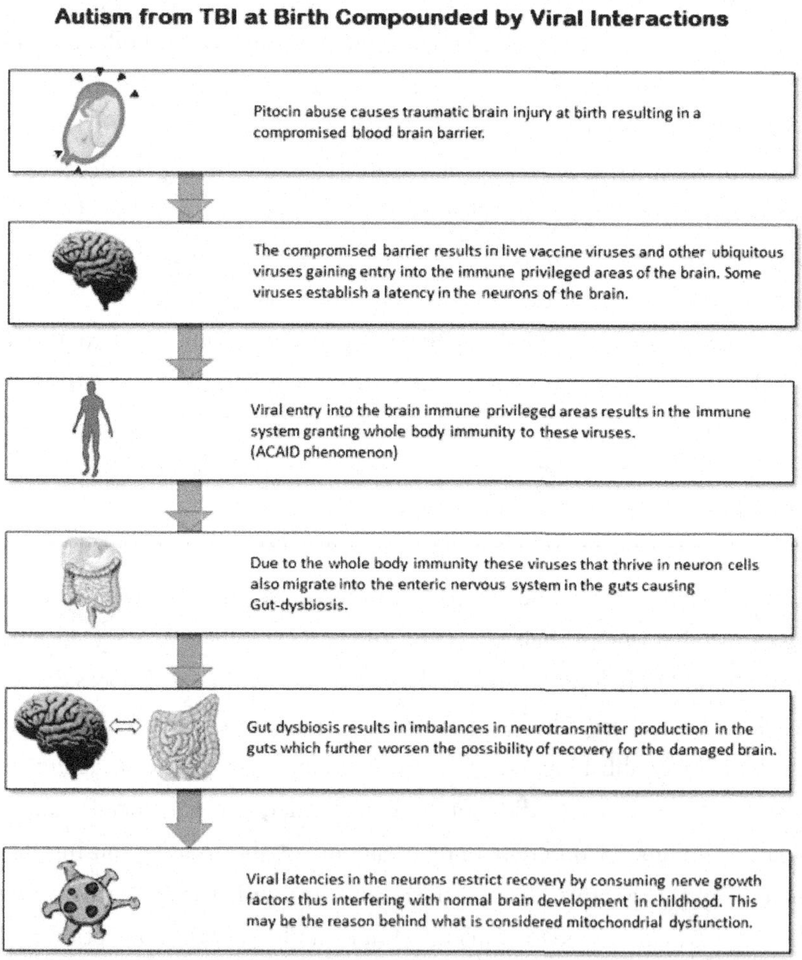

Fig 9.2: Viral interaction with pre-existing brain injury, causing gut-dysbiosis.

Thus, live vaccine viruses, inflammatory responses and antibody interference might actually have an impact on early brain development, particularly where a previously damaged brain is involved. This explains why only a small fraction of parents observe retrogression of skills in their kids after vaccination and why only certain people are impacted (and not everybody reacts the same way to vaccination). These impacted kids might have recuperated from their brain damage had they not been subjected to a constant barrage of viral vaccines in early childhood, coinciding with the time of maximum brain development. This period is also the window during which the maximum possibility for recovery from brain injury exists. Thus, in cases of pre-existing brain injury vaccines actually represent the second blow to a developing brain causing the autism epidemic.

Diseases, Vaccines and Allergies

Most vaccines come with warnings of allergies including rare occurrence of anaphylactic shock. The science behind immunity, vaccines and their interactions with our bodies is very complex, but the logical explanation to vaccine related allergy effects is actually straightforward. Some viruses pass through our bodies without causing disease. They lack the right protein structure to invade our body cells and trigger an immune response. Some others evade the immune system by staying benign. Then some others have proteins that help them invade certain body cells and replicate out of control triggering disease and an immune response.

Some viruses are eliminated from the body by the immune system, and some others continue to hide in our bodies establishing episomal and proviral latencies in certain cells of our bodies as described previously. In cases where the viruses are completely eliminated from the body, the immune memory doesn't typically last too long. That's simply the way the human body is designed and the way the immune system works. A long term continuous inflammatory immune response is detrimental to the body and the immune system itself. Once the invading antigen is subdued, the immune response is scaled back and the activated immune system becomes inactive with innate immunity taking over.

Some diseases like measles and chicken pox provide lifelong immunity, simply because the virus becomes latent and continues to hide in certain parts of the body and have to be continuously kept under check by the immune system. The immune system achieves this by maintaining a high level of antibodies to the disease at all times. This maintenance of higher levels of antibodies is the reason for long term allergies. Shingles for example is a disease caused by the re-activation of the chicken pox (varicella) virus later in life when the immune system loses its grip on the virus.

My observation has been that diseases which produce long term immunity also cause allergies that stay for the life of the individual. Due to their establishing latency in the body and the need for long term suppression, the affected persons immune system itself gets changed in ways that result in a lifetime of allergies. Thus, long term immunity equates to a lifetime of allergies. If such allergies happen due to viral latencies in the neurons of the brain and the enteric neurons of the gut, due to pre-existing brain damage, the result can be an exacerbation of autism symptoms and prevention of damage healing.

The deal with vaccines is that vaccine manufacturers, health officials and everybody else for that matter, want vaccines to provide lifelong immunity. Nobody wants to be getting frequent vaccines (getting frequent vaccines itself has been shown to be counterproductive). Long term efficacy is actually difficult to create, so manufacturers add numerous other ingredients to the vaccines to achieve a high level of immune response and that long term immunity. Unfortunately, such immunity is achieved from a constant high level of antibodies and the result is the accompanying allergies and inflammatory diseases.

What Should Parents Do?

The compulsions of modern society where mothers are part of the workforce also means newborn infants and babies are forced into daycare where they can unwittingly spread diseases to epidemic proportions. The modern answer to this problem has been to vaccinate kids. The sheer number of diseases and vaccines a current-day human infant will be vaccinated against has grown 10-fold since the 80's. While any disease left undiagnosed, unsupported or untreated can turn deadly, it has become a matter of schedule and convenience to vaccinate. But with the vaccine autism relationship making headlines some parents have begun making a choice not to vaccinate. The truth remains that nobody wants to become a part of the minority statistic of where a disease turned deadly or end up with a lifelong debilitation for their child. Most parents are caught up in the conflicting information about the vaccine autism relationship and the need to protect their babies from debilitating and life-threatening diseases.

Modern hygiene standards and family circumstances can allow some of these vaccines to be delayed. In cases where a mom, dad or other caretaker are looking after a new born at home, it may be possible to work with a pediatrician to delay some of the live virus vaccines. However, if a child is going to be in a day care situation, parents' options may be limited, which is where their knowledge and commonsense becomes paramount. The vaccine schedule is made with the earliest possible vaccine administration ages in mind. The Center for Disease Control (CDC) publishes a vaccination schedule and also possible contra-indications to vaccines. It is advisable for parents to go through the schedule, study the contra-indications, identify the live vaccines in there and figure out the best course of action for them with their pediatrician with due consideration of the birth circumstances. The caveat is most pediatricians are trained about the positives of vaccination and not the other side of the story. You may have to interview them for their point of view.

The real risk and most important factor with the vaccine decision for your child is the potential for reinjuring their brains due to the live virus found in many vaccines. Some criteria for new parents looking to vaccinate your babies is to look back at your birth experience for possibilities of brain damage:

- Were you induced for more than 4-5 hours?
- Did you have severe blood pressure/ gestational diabetes issues?
- Was the baby premature for any reason where doctors suspected neuro issues?

- Did the mother have infections/fevers during pregnancy?
- Were you C-sectioned due to fetal distress of any kind?

While not be an exhaustive list, if you answered yes to any of these questions, then it is possible your child has brain damage. As a parent you know if you had a traumatic birth for any reason, in which case it might make sense to wait on the live vaccines and any other vaccines like DTP with listed neurologic side effects until after age 6, by which time brain development is past the phase of maximum development. By then you would know for sure if the child has recovered from any possible brain injury sustained at birth. If a child has development issues past age 6 in spite of not being subjected to live vaccines, you might consider keeping him or her away from vaccines for the rest of their lives and hope they get 'herd protection' from these viruses by way of healthy individuals around them being all vaccinated.

Parents and pediatricians need to be aware of the arguments for and against vaccinations and make an informed decision on a case-by-case basis. There is absolutely no reason for parents that had a natural, trouble-free pregnancy and birth to keep their kids away from vaccines. In their case the effects of getting the disease are far worse than the effects of the vaccines. Indeed, it's the responsibility of those fortunate parents who had a problem-free delivery or had expert OB/GYNs that helped navigate their pregnancy and delivery without brain damage to vaccinate their kids so as to offer herd protection to the unfortunate ones that suffered brain damage at birth.

Key Take Away: While vaccines may not be directly responsible for autism, when pre-existing brain damage exists, vaccines and their interactions with a previously damaged brain results in further reduction in brain capacity that is noticed by parents as retrogression in development. This is particularly true of live vaccines like measles where the viruses create a neuronal viral latency. Pediatricians, medical providers and health officials need to stop being obstinate about vaccinating all kids and instead take into consideration birth trauma circumstances and (at the least) avoid live vaccines for such high-risk kids.

Part IV
The Learning

In this section we learn about the four different "tracks" of complications which are the true causes leading to autism and related disorders. We also investigate the possible treatment options based on our understanding of the causes. While a complete recovery may not be possible, with the current state of medical technology, it is important to foster and advocate for new technologies that pursue the possibility of a full recovery.

CHAPTER 10

Pathways to Developmental Disorders

In talking to other impacted parents over the years and based on my own knowledge of what happened to my boys, I have been able to draw conclusions on the different causative pathways to autism and other neurological disorders. I have put together my conclusions based on some probing questions I had asked other parents and their descriptions of their circumstances and experience from pregnancy through birth and beyond. The actual percentage of cases in each pathway is based on my small sampling of data, so that could vary if studied at a more regional or national level, although it is probably still ballpark. The more I have investigated these questions, the more I realize the answers can only come from impacted parents. Research studies are typically focused on specific causative factors while (as you will see here) the causes are just as much a broad spectrum of reasons as the consequences are. These causes and diagnosis can come over prolonged periods of time, which can only be pinned down by parents and caregivers associated with the impacted individual over the entire period. Medical malpractice also plays a significant role, something academic research never wants to touch with a 10-foot pole.

I have identified the following four causative pathways into which I can categorize most of the causes of neurodevelopmental challenges in children. Three of these lead to the condition of autism.

- Non-Autistic Genetic Disorders (10%)
- Maternal complications during pregnancy (10%)
- Natural delivery trauma at birth (30%)
- Traumatic Brain Injury at birth caused by malpractice (50%)

While there is no preventing genetic disorders, the others are causes that expert OB/GYNs can help their patients successfully navigate through with proper pregnancy and obstetrics care; indeed, minimizing these risks is one of the reasons people go to hospitals for birth in the first place. Figure 10.1 summarizes the four major neuro disorder tracks.

The DSM definitions of the autism spectrum have changed over time and they are now in their 5th iteration. I believe this is due to trying to define it from the symptomatic side instead of trying to name it from the root cause side of the issue. This kind of root cause diagnosis-based naming may help doctors identify and treat the causes of the symptoms better, resulting in earlier intervention and better outcomes for the impacted individuals. For example, symptoms are slightly different based on the circumstances causing the autism. Boys with traumatic brain injury from Pitocin abuse tend to have more behavior issues (e.g. flapping and visual and vocal stimming) compared to boys with hypoxic damage from cord complications. Boys with hypoxic damage tend to have more memory, learning, anxiety, anger control issues, while the problem with boys with traumatic brain injury tends to be one of hyperactivity and lack of focus more than memory or ability to learn. Traumatic brain injury tends to impact language ability a lot more than hypoxic damage. Calling it "PDD-NOS" (Pervasive development disorder not otherwise specified) and rolling all these symptoms into the autism spectrum disorder as in the DSM–5 only adds to confusion rather than clarify the disease condition.

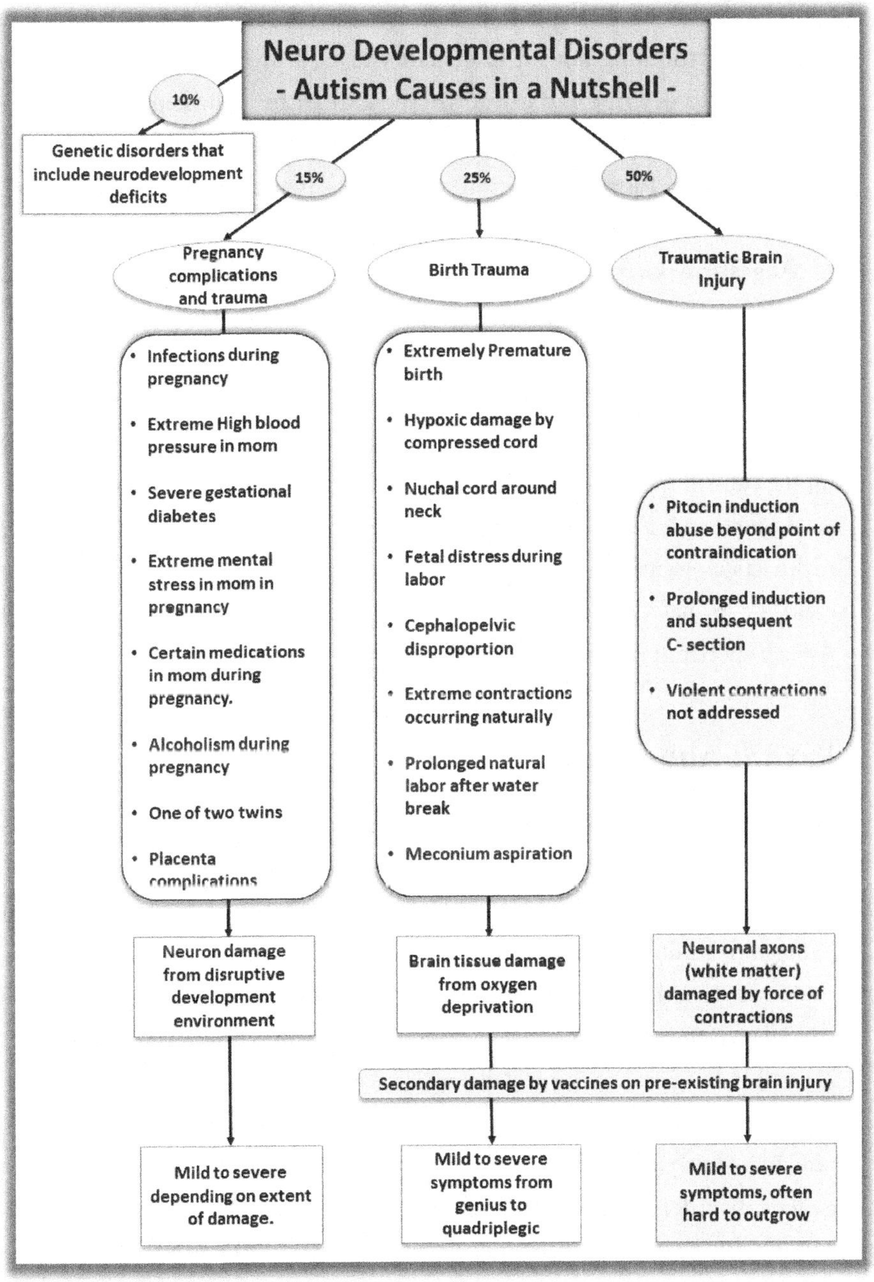

Fig. 10.1: Illustration of the various tracks.

In order to better explain the causes of autism and remove confusions related to attributing autism to genetic causes, I propose the following four new words to describe and categorize the four paths identified above:

Augenism: Genetic mutations causing neurodevelopmental disorders in a child. A more formal definition would be "neurodevelopmental disorder traceable to genetic deficiencies."

Augestism: Gestational or pregnancy complications creating developmental disorders in a child. A more formal definition would be "a neurodevelopmental disorder on the Autism spectrum traceable to complications during pregnancy."

Aupartism: Birth (parturition) complications creating developmental disorders in a child. A formal definition would be "a neurodevelopmental disorder on the Autism spectrum traceable to complications during parturition."

Aumalism/ Autism Iatrogenic: Perinatal (at or near birth) medical malpractice or iatrogenic brain injury resulting in developmental disorders in a child. A formal definition would be "a neurodevelopmental disorder in the Autism spectrum traceable to iatrogenic injury in the delivery of the individual."

Path 1: Augenism

Genetic causes are usually beyond the control of people and science and a matter of destiny. They are known to happen as a result of in-breeding within a genetic pool. Correlations have been established, which point to genetic damage from environmental chemicals, viruses or age of parents. Sometimes there are genetic traits that run in families. But modern science can find specialized cures in some cases including gene therapies where the cause is identified by gene mapping deficiencies within the genome of the impacted individual. The percentage of these cases among the total population have remained fairly small and very constant over time. Only a small subset of genetic disorders cause neurodevelopmental deficits. Their symptoms tend to be very different from autism, but they are often confused with Autism due to similarities in how they impact neurodevelopment.

Down's Syndrome: This is the most common genetic abnormality with mental health implications and occurs in about 1 in 1000 individuals.

Normally people have 23 pairs of chromosomes in the nucleus of their body cells. Down's syndrome happens when an extra full or partial copy of chromosome 21 appears in the cells of the person due to the genetic copy process going wrong at conception. It causes distinctive features in the eyes and nose, smaller ears, hands, legs and other conditions affecting mostly hearing, the heart, and sleep. Their health challenges can sometimes keep them from living a long life and they sometimes have learning difficulties, but children with the condition are capable of speech and even achieve limited independence in adulthood. Its occurrence has been correlated to the age of the mother. However, the same mother can have subsequent babies that don't have the syndrome, which makes it despite such correlation a pure game of chance. Down's syndrome is not autism but approximately 20% of kids with it are also said to have an autism diagnosis.

Screening for presence of Downs syndrome during pregnancy is now available and utilized almost routinely along with screening for other birth defects in many parts of the world. At about 10-14 weeks of pregnancy an ultrasound to detect fluid collection at the back of the baby's neck, combined with maternal blood testing is used as an indicator. Genetic testing of amniotic fluid in the 15-20 weeks of pregnancy using a procedure called amniocentesis can confirm the presence of this syndrome and others. Parents mostly opt to abort the pregnancy if such a defect is detected and this has resulted in a ethical dilemma and debate in many countries. Some other countries have claimed total elimination of Downs syndrome births as a result of such testing.

Craniosynostosis: The human skull (cranium) is composed of 22 different bones encasing the brain and forming the facial features. Of these, 8 bones are cranial bone plates that encase the brain. These are held together by unique joints called sutures, which are made of thick connective tissue. They're irregularly shaped, allowing them to tightly join all the uniquely shaped cranial bones. The sutures don't fuse until adulthood, which allows your brain to continue growing during childhood and adolescence. Craniosynostosis is a malformation that involves the early closure of one or more sutures of the skull, usually prior to birth. Craniosynostosis may occur in isolation or this condition may be associated with other abnormalities as part of a syndrome.

There are close to 200 known genetic disorders which include craniosynostosis as one of the effects. These syndromes are primarily differentiated by the type of suture and gene mutation involved. One in 2500 births are known to involve craniosynostosis. Natural birth requires that these bones be separable so that they can take the right shape to help the baby make it safely through the birth canal without injury to the brain. Fused cranial plates at birth can result in brain injury through crushing and deformation of the head. Some of the underlying conditions that cause craniosynostosis can also cause neurodevelopmental disorders.

Dyslexia: Dyslexia is a learning disorder that involves difficulty reading due to problems identifying speech sounds and learning how they relate to letters and words (decoding). Also called reading disability, dyslexia affects areas of the brain that process language. It tends to run in families which is why it is categorized as a genetic disorder. Studies supported by the UK MRC, the Wellcome Trust, Yale and the NIH have identified a few genes responsible for the learning disability. Just as in autism it can involve speech deficiencies in addition to reading. However, speech difficulty in autism tends to be due to lack of focus and slow processing speed rather than underlying language reading abilities. It is said that about 1 in 4500 newborns is diagnosed with dyslexia.

There are over 250 other genetic defects like fragile X syndrome that are very rare, each with its own specific set of symptoms, some including impacts to brain development. Underlying genetic conditions can also cause several neurodegenerative and metabolic conditions that slowly progress over time. One final (though very rare) genetic disorder that merits mention here simply because the kind of damage can look very similar to TBI is *Microcephaly-capillary malformation (MIC-CAP) syndrome.* In cases of medically induced TBI at birth medical professionals tend to blame this syndrome to escape culpability. The MIC-CAP syndrome is caused by a mutation in the STAMBP gene causing microcephaly, an abnormally small head resulting from problems with brain development during pregnancy.

Mutations to this gene can result in cells dying during development, affecting the entire development process.

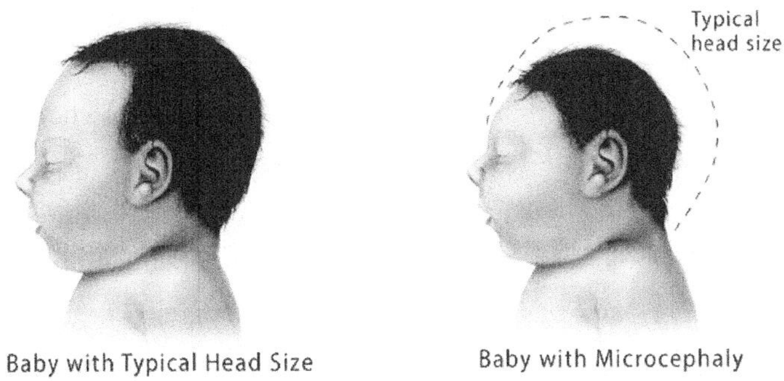

Fig. 10.2: Difference in head sizes normal vs microcephaly.

But microcephaly is distinct from autism or any of the other typical developmental disorders. Kids with autism have normal head sizes and exhibit no other bodily disorders except those stemming from brain damage and related malfunctioning.

Maternal genetics vis-à-vis pregnancy: This is the only real genetic issue that is fairly widespread and is a cause of autism symptoms in some children. This has to do with the ability of the mother to support the baby in the third trimester of pregnancy. Some mothers have traits running in the family where they have a tendency to lose the fetus late in the pregnancy. The exact causes are unclear, but it could be due to various circumstances mostly related to placental issues (and brain damage is typically caused by insufficient oxygen that is supplied through the placenta and the umbilical cord). This typically results in the death of the baby, but in the rare occasion that this is caught ahead of time the result is a baby with autism.

Path 2: Augestism

Maternal Complications: It is traditional wisdom and common knowledge that pregnant women need to be well cared for, with good food and emotional support during her pregnancy to ensure the best outcomes for the baby.

Recent studies in the field of epigenetics suggest that a mother's experiences and habits during pregnancy can play a role in activating or de-activating genes in the fetus, which can play a significant role in the future life of the baby. A lot of these changes show up in the long term and do not impact the immediate birth outcome significantly. This is also the reason modern medicine recommends prenatal vitamins to ensure there are no chances of vitamin deficiencies in mothers as these can impact fetal development.

A pregnant woman's body undergoes a lot of changes as it prepares for the nourishment, growth and delivery of a baby. Some pregnancy related changes in the mother's body, like developing high blood pressure or gestational diabetes or both can happen and they can have impacts on the fetus and birth. In most cases, the impacts of maternal complications are small, and a lot of times impacted kids are able to speak, but have other characteristics like social anxiety, emotional extremes and such. These are diagnosed as higher functioning autism, but in worst case scenarios when the mother contracts diseases such as cytomegalovirus infection, herpes simplex virus infection, rubella, hepatitis B, syphilis or some food borne illnesses such as listeriosis or toxoplasmosis, it can result in serious negative outcomes for the baby including mental retardation, microcephaly, cerebral palsy, or even stillbirths. Zika is one such new virus that made headlines for the impact it had on brain growth in babies of pregnant mothers who contracted the illness. When there is no occurrence of any of these effects, maternal complications mostly result in milder forms of development delays which can be treated very well with early intervention programs.

High blood pressure: High blood pressure during pregnancy if not stress-related is mostly an attempt by the mother's body to ensure the fetus is adequately supplied with oxygenated blood and nutrients. It can be a reflection of how strongly the placenta is attached to the blood supply and how well supplied it really is. As the baby grows bigger, the demands on the mother's body to keep it supplied with nutrients increases as well. If the placenta for some reason is not tapped well enough into the mother's blood supply through the uterus then it can result in the body trying hard to compensate by a higher blood pressure. Thus, high blood pressure can be an indication the fetus may not be getting enough oxygen and nutrients flowing through. It is particularly common toward the 3rd trimester, and although most babies do not have any complications from this, in the most seriously impacted (where the blood reaching the fetal brain is not sufficient) the results can be autism symptoms.

Gestational diabetes in mother: Higher blood sugar levels in mothers means the baby's pancreas has to work harder to control its sugar levels. The higher sugar levels also mean higher fat storage in the baby, making for bigger babies. Such babies are at risk of hypoglycemia when that sugar level drops drastically after birth as the pancreas adjusts to the more normal levels. A bigger baby can also lead to birth complications in vaginal delivery. When the mother takes medications to keep her blood sugar under control, the impacts of the medication on the baby are not clearly understood. There can be additional effects caused by using sugar substitutes: for example, aspartame is a neurotransmitter. When used by the mother as a sweetener to replace sugar to control her gestational diabetes, it could impact the fetus by neurotransmitter overdose (excitotoxicity) causing neuron apoptosis (suicide) or possibly inhibit neurogenesis in the fetal brain.

Although I do not have a lot of data to prove it, another interesting outcome that I have sometimes observed is differences in birth when the mother has gestational diabetes: C-section babies turn out fine while babies delivered vaginally develop autism. Perhaps the diabetes and high sugar levels in the mother make the neurons in the baby's brain brittle or in some other way more susceptible to damage when subject to the pressures of vaginal delivery.

A skilled obstetrician can help mothers navigate this risk through managing the diabetes with restrictive diets and food regimens, controlling the natural weight gain in pregnancy and lifestyle changes which can help control diabetes without medications and result in a healthy baby.

Obesity in mother: This is a tough one to single out, simply because mothers that are very obese prior to pregnancy tend to have a lot of the other complications described here during pregnancy and it is actually very difficult to ascertain the true cause of the child's complications. But regardless of whatever complications in the mother's health that might be causing the mother's obesity, like hormonal imbalances, metabolic disorders, mental challenges, etc., these can also cause elevated blood pressure, gestational diabetes, etc. However, at the hands of an exceptionally capable obstetrician many such mothers can navigate their pregnancy very successfully and deliver healthy normal babies in many cases.

Path 3: Aupartism

Birth (parturition) trauma, particularly oxygen deprivation (asphyxia) due to factors ranging from cord prolapse to complications of twin birth ranks as the second-highest cause of autism. The lifelong neurological consequences of in-utero asphyxia are very well known to the medical community and very well documented. These complications that happen naturally are the primary reason why a hospital birth with careful monitoring of the fetus is warranted in the first place. Typically, outside of accidental complications like nuchal cords described below, most mothers that have maintained a healthy lifestyle prior to and during pregnancy tend not to run into most of these complications in their babies.

Extremely premature birth: Babies born very premature (before 37 weeks of pregnancy) do not have a fully functional body ready to take on the challenges of the world. When they leave the protection of the mother's womb ahead of time, their lungs are not fully developed. This can lead to deficiencies in their vital oxygen supply necessary to sustain the rest of the body and can have permanent adverse effects on the future of the child. The more premature a baby the more the possibility of permanent brain damage. As the organ consuming the most oxygen, the brain is the most vulnerable to hypoxic damage.

Umbilical Cord Issues: Umbilical cord issues tend to be more common than they're given credit for, and when they do not cause the stillbirth of the baby, they contribute to a significant fraction of autism impacted kids.

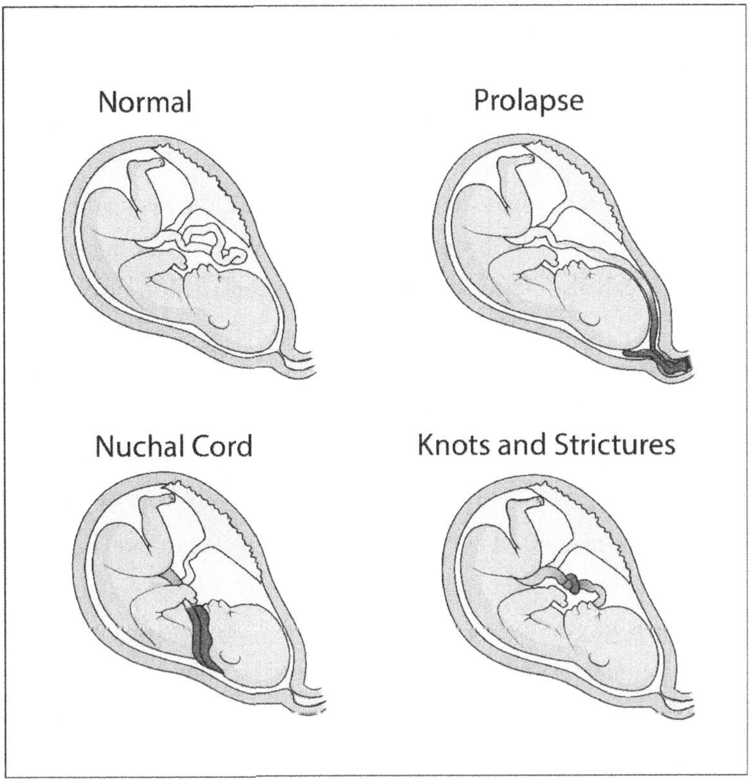

Fig. 10.3: Umbilical Cord Issues

The fetus derives its oxygen and nutrients from its mother while in the womb. A compressed cord most often happens when the cord is prolapsed or stuck in the birth canal ahead of the baby. When this happens, it can cause compression in the cord, cutting the flow of vital oxygen to the baby, which if left untreated can result in hypoxic brain damage and even death of the baby in what is called a cord accident

Nuchal cords happen when the baby gets tangled in the umbilical cord while swimming in the womb. The umbilical cord is 20-24 inches long typically, but long cords of 32 inches or more are known to occur. The longer cords can tangle around the baby and end up looping around the neck. They typically untangle by themselves in the buoyant fluid-filled environment of the womb. When the umbilical cord knots around itself or gets kinked in some very rare instances, it can again restrict blood flow and gas exchange between the mother and fetus, resulting in hypoxic brain injury.

Since the baby is not breathing through the lungs yet, a cord around the neck is not usually a problem. But the problem occurs when the cord tightens and compresses to a point where the oxygen flow through the cord is impacted, particularly when the cord is still around the neck while the baby makes its way through the birth canal. The blood flow through the neck to the fetal brain can also be constricted and such constriction of blood flow to any part of the body is called ischemia. Thus, an injury involving both lack of oxygen in the blood itself (hypoxia) and lack of blood flow (ischemia) leads to what is called hypoxic-ischemic injury. This happens in less than 2% of births, but when it happens, it can result in stillbirths and babies that are born with an extreme blue or dark skin tone due to lack of oxygen. Boys in particular tend to be later diagnosed with autism. Cord accidents are known to be responsible for about 10% of all stillbirths.

Fetal distress in late stage pregnancy and labor: As the baby grows bigger, more nutrition and oxygen are needed to keep the baby healthy. While it is not a problem in most cases, sometimes other maternal complications like high blood pressure, placenta complications, external trauma or maternal habits might result in the fetus not getting sufficient oxygen leading to distress which might result in a doctor recommending an emergency C-section during one of the prenatal checkups. Such kids born from distress might suffer neurodevelopmental delays.

Placental abruption: This is the placenta detaching from uterus before baby is born. While this is extremely rare, it is known to happen where the placenta partially detaches causing internal or external bleeding. It can happen anytime during pregnancy, but the propensity is more in the perinatal stages.

While it is known to occur only less than 1% of cases, it can cause death of the baby from insufficient oxygen and nutrient supply. In milder cases it can lead to growth restriction, fetal distress, acidosis and lead to brain damage in the baby.

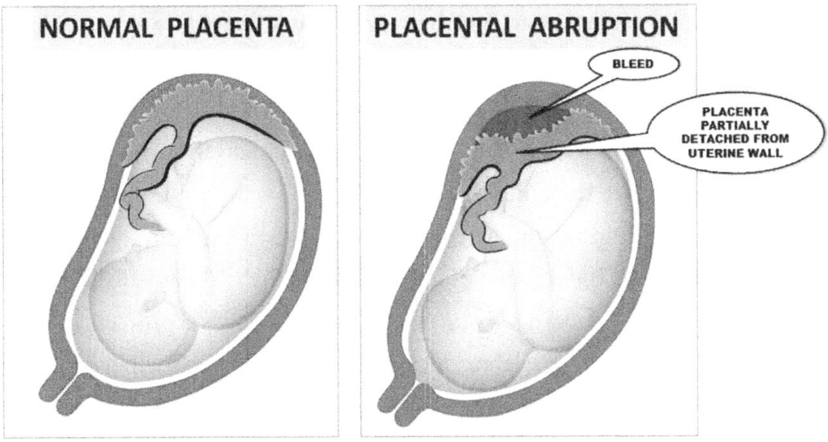

Fig. 10.4: Normal placenta vs abruption.

Ductus arteriosus premature closure: A special blood vessel in fetuses helps bypass the nonfunctioning lungs before birth. This is supposed to close after birth and become a remnant ligament. But premature closure is known to occur mainly as a side effect of maternal medications. A premature partial closure affects the heart, circulation and oxygenation with adverse effects.

Cholestasis or jaundice as a result of pregnancy: Blocking of the bile duct in the mother due to side effects of medication or other causes results in jaundice in mother and baby during pregnancy. This condition is associated with an increased risk of premature delivery and even stillbirth. Some infants born to mothers with intrahepatic cholestasis have a slow heart rate and suffer from a lack of oxygen during delivery, leading to fetal distress and all the other complications that arise from it including neurodevelopmental disorders.

Cephalopelvic disproportion (CPD): This occurs when the head of the baby is considered too large to fit in through the mother's pelvis. Both the pelvic bone structure and the head of the baby are designed to give. Even with a 10cm dilation for the head to emerge from the cervix and allowing for some skull coning, there is still obviously a minimum pelvic diameter necessary for a baby to pass through, and at times there are still complications. CPD is a problem that's been neglected by modern obstetrics and remains one of the most frequent reasons for C-sections, but it also presents one of the most common risks of brain damage to a baby when doctors try for vaginal delivery.

Prolonged natural labor after water break: Doctors disagree on how long after a woman's water breaks that a baby needs to be delivered for achieving best health outcomes for the baby. Recommendations vary from about 12 to about 72 hours after the amnion ruptures for the baby to be delivered. The lack of amniotic fluid heightens certain other risks if the baby doesn't come soon and labor gets stuck. Some of the risks to the baby include infection and a higher possibility of cord compression from the lack of protection of the amniotic fluid. Prolonged labor may cause one of these other situations leading to hypoxic brain damage or could result in traumatic damage from the force of uterine contractions.

Meconium Aspiration Syndrome (MAS): Meconium is a dark green sticky substance that accumulates in the intestines of a fetus, comprised of gastrointestinal secretions and other waste accumulated mostly during the third trimester from its limited metabolic functions. Meconium is passed typically within 24 to 48 hours of birth under normal circumstances. However, some fetuses tend to pass the meconium while still in the fluid-filled amniotic sac, which is suspected to potentially cause fetal distress due to hypoxia. In about 5% of fetuses exposed to meconium, when the fluid-filled lungs of the fetus transition to the air breathing organs of the child difficulties emerge due to factors like blocked air passages, inflammation, infection, and lung surfactant inactivation. Development of MAS as a neonate has been shown to be related to longer-term neurodevelopmental disorders such as autism.

Multiple births (twins or more): Twin births can add an additional layer of complexity to the already intricate and complex pregnancy process. Dizygotic or fraternal twins are formed from 2 eggs and embryos and always have separate placentas and amniotic sacs. In the case of monozygotic twins or multiples where a single egg splits after a few days after conception, they might come with shared or separate placentas and share the chorion or the outer layer of the amniotic sac while have their own amnion or internal layer of the fluid-filled amniotic sac. Most twins or multiples come with their own placentas, their own amniotic sacs and umbilical cords. Monoamniotic-monochorionic (MoMo) twins with shared placentas chorion and amniotic sacs occur in less than 1 percent of twin births. MoMo twins are at extremely high risk for cord entanglement, cord compression, and other complications that can lead to autism.

Navigating these complexities skillfully are what OB/GYN's are trained to do, with full awareness of the consequences of not making timely decisions. These are the reasons why we go to the hospital for birthing in the first place. Most OB/GYNs tend to make the right decisions and are trained to save the life of the baby, but sometimes their inaction can result in loss of valuable time and the health of the baby suffers as a result, even when the baby is delivered live. The mental health of babies is something many OB/GYNs seem to take lightly, using neuroplasticity as the excuse for allowing the brains of babies to take a beating during birth. Consequently perinatal malpractice has become fairly rampant.

Path 4: Aumalism / Autism Iatrogenic

Pitocin abuse: This path to autism represents the highest number of autism cases out there. I estimate at least about 50% of the cases of autism in the US are due to malpractice of this variety. Indeed, all the prior reasons, risks and pathways have existed for centuries and cannot explain the recent meteoric rise of the number of cases of autism. This is the single reason why autism is so much more prevalent in western countries and also the reason why autism is on the rise.

The dangers of Pitocin are described right on the label of this drug. It has become the most abused medication in maternity wards across the country. Unfortunately, there's a lot of people including health care professionals who do not understand the induced labor process as much as they should. When a drug becomes so common, it is easy to let one's guard down and start treating it as the 'safest' thing out there, forgetting the warnings on the label. Unfortunately for parents and their kids impacted from Pitocin, the other routes to autism obscure the reality of the prevalence of this pathway.

The single most important line on the label about the circumstances for contra-indications of Pitocin use that OB/GYNs don't pay attention to is this:

5. Where adequate uterine activity fails to achieve satisfactory progress.

"Satisfactory progress" is subject to interpretation. Most maternity wards interpret it as about 4-6 hours of Pitocin induction at a low dosage without any progress in cervical dilation. What it doesn't say is what happens if you cross that limit. In many cases, in the absence of any other adverse reactions like fetal distress, some OB/GYNs have resorted to increasing the Pitocin dosage instead of going ahead with a C-section or at least stopping the induction.

Increasing Pitocin dosage at this point is a sure shot way to damage the fetal brain through physical trauma that leads to scarring in the white matter of the brain – the fundamental networks of neurons that become the foundation that the future neuronal network will be built on. And unlike an adverse chemical reaction, physical trauma is guaranteed to happen to any fetus subject to Pitocin abuse.

Indeed, this is almost the only situation where I have seen girls develop autism (although their numbers are fewer due to their ability of recover from the injury faster due to the brain protective nature of estrogen). It's not uncommon to find girls with scars in the brain who otherwise seem to be normal. It is also common to find the same mother induced the same way have one boy on the autism spectrum and a girl that might be fairly normal.

What's most unfortunate about this condition is that these OB/GYNs proceed with induction in the absence of other risk factors like fetal distress. The impacted individuals are the healthiest mothers and healthiest babies who end up paying a penalty for their good health at the hands of these OB/GYNs and their Pitocin drug abuse.

Almost every article you find on the internet from doctors and researchers out there about the Pitocin – autism linkage talks about the chemical effects of the oxytocin hormone on the fetal brain as somehow being the explanation for autism. What they don't talk about is the sheer trauma that artificially induced contractions can produce on the fetal brain causing traumatic brain injury. It's not a chemical adverse reaction but the effect of the physics of the sheer force of artificial contractions hammering an infant head first on a non-effacing cervix.

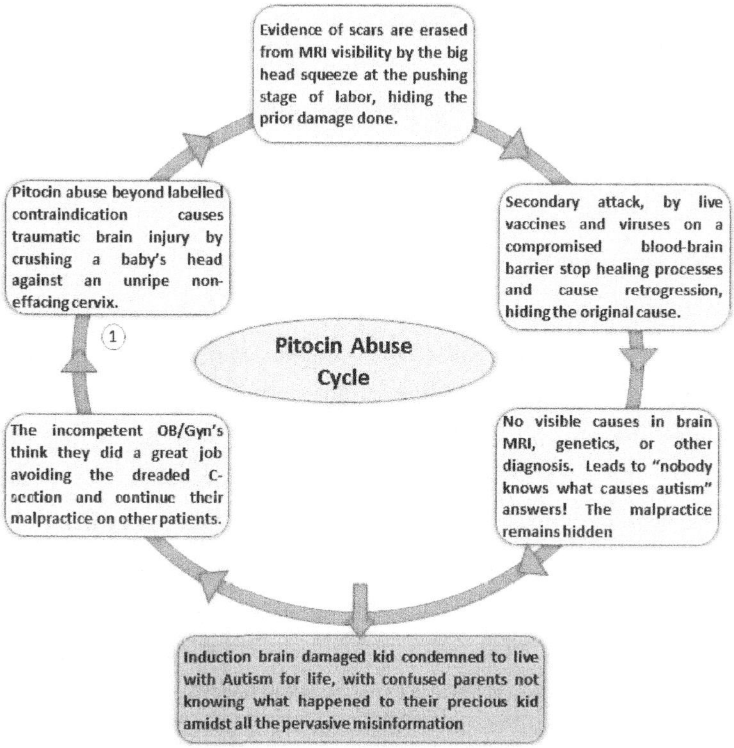

Fig. 10.5: Pitocin Abuse Cycle

Some conscientious researchers and OB/GYNs have tried to critique the practice in their hospitals. One such paper written by British researchers Karl Oláh and Phillip Steer is titled "The use and abuse of oxytocin" in the Journal of the Royal College of OB/GYNs[10.2] and goes into detail of how nurses do not understand the terminology involved in induced labor and the mistakes and misinterpretation in cardiotocography. They elaborate on the methods, intricacies and outcomes of asphyxia in many of the deliveries when strict guidelines and careful monitoring is not practiced. But for the most part the issue has been ignored as it falls under 'medical malpractice' and would break the taboo of doctors blaming other doctors and exposing their malpractice.

This has caused the burden of proof to shift squarely onto the shoulders of impacted parents.

Aumalism or Iatrogenic autism can be eliminated once the recognition of Pitocin abuse gains common acceptance and interventions that can cause traumatic brain injury are avoided. Aupartism can be effectively reduced through better education of mothers and caregivers to watch out for the possibilities.

Key Take Away: Four different causative pathways to autism are catalogued and a clearer labeling of autism subcategories is proposed, distinguishing autism from genetic disorders and establishing its roots in pregnancy and birth. Adopting such categorization could lead to better understanding of an individual's condition, better categorization of symptoms and more directed treatments, leading to a significant reduction in Autism overall. The complications of birth are discussed and the role of perinatal medical malpractice in contributing a significant portion of the autism epidemic is clearly identified.

CHAPTER 11

Treatment Possibilities.

Impacted parents are desperate enough and pushed to a limit where they can fall prey to buying into anything that is touted as a possible cure. There are numerous websites with testimonials from people touting false promises. The internet can be a dangerous place where parents who are not discerning enough stand to lose thousands in hard-earned money and valuable time pursuing a path that bears no results and sometimes might actually cause more harm or damage their kid even further.

While some cases of these testimonials may be true, they are potentially one-off cases where the cause for the autism symptoms may not have been traumatic brain injury or hypoxia but one of the causes curable by other means such as Pandas, pans, Lyme disease or just an allergy that needed to be controlled. Unfortunately, medical science falls short when it comes to a cure to autism caused by TBI and hypoxia. In a cruel world where medical sciences bluntly tell parents "there's no cure for autism," hope turns to desperation. It's the last thing any impacted parent wants to hear. They would rather hold onto hope that their kids will be better one day and work toward it. As the character of Tim Robbins in the movie *The Shawshank Redemption* says, "Hope is a good thing, maybe the best of things, and no good thing ever dies."

The Damaged Brain

We know that brain damage at birth is the cause of neurodevelopmental delays including autism, ADD, ADHD, etc. Consider Fig 11.1a and 11.1b showing fetal brain development and a picture of a child's brain in cross-section vs a young adult brain. The biggest physical change is the increase in folds of the brain. At 9 months from conception close to birth, the brain has already developed the basic folded cerebral structure. With age cerebral ridges become more pronounced as the development of new skills requires growth of more grey matter within the limited cranial space. More folds help accommodate the increased volume of grey matter on the cerebral surface to house all those additional neuron bodies that make up the grey matter in a process called gyrification. As the brain undergoes more experiential learning with age, more grey matter develops to store all those additional skills and information.

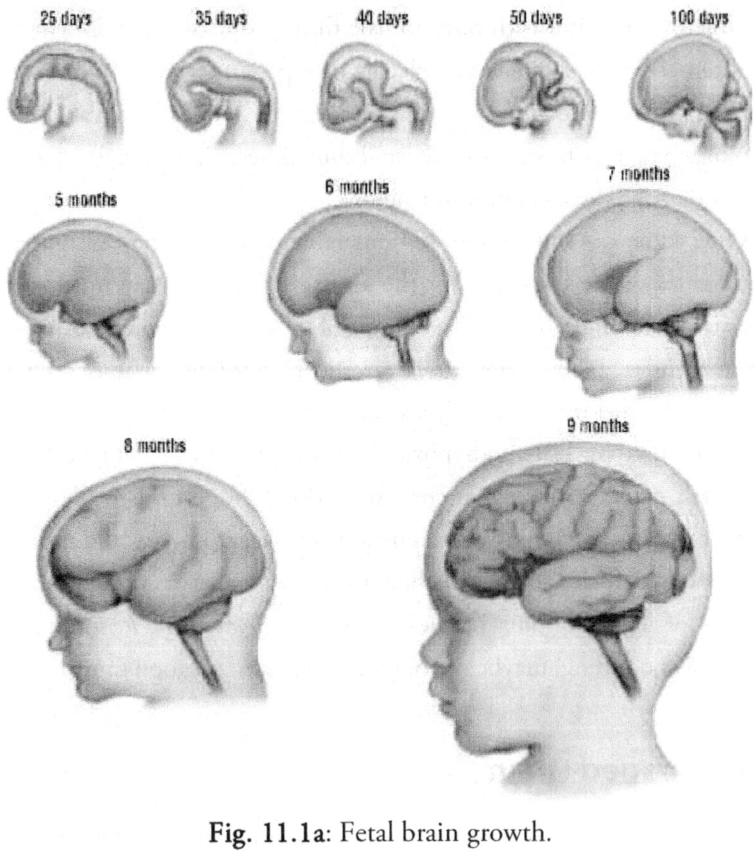

Fig. 11.1a: Fetal brain growth.

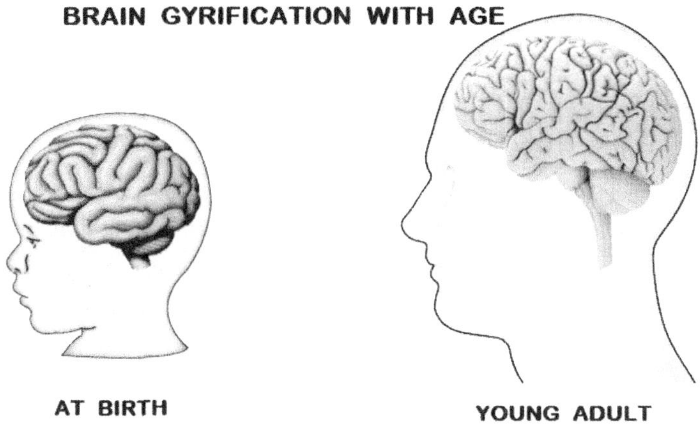

Fig 11.1b: Increased brain folds with age.

As seen in the pictures during fetal development and from birth through adulthood this increase in ridges on the brain surface is what happens in normal brain development, in a process called gyrification as the brain makes more surface area to accommodate more learnt skills.

Coordination of multiple grey matter areas involves the neural connections in the white matter, which connect the different areas of grey matter. For example, speech involves auditory processing, visual processing, cognition and memory – each of which by itself involves multiple areas of grey matter, each working with other areas it is connected to by white matter.

In a damaged brain such experiential learning is pretty restricted. This is because the foundational networks that connect the visual, auditory, memory and sensory processing are damaged, resulting in the grey matter areas needed to participate in an activity not communicating correctly. Worldly viruses crossing the compromised blood brain barrier create further imbalances in brain signaling and neuron function.

Autism resulting from brain damage to either the grey matter or white matter areas (or both) followed by a double whammy from live viruses gaining control of immune privileged areas, means that any cure needs to address the remediation of such damage:

- Help repair and recover from brain trauma;
- Help regrowth of neurons;
- Help with the removal of scar tissue;
- Help re-establish damaged connections between neurons and create new pathways;
- Reduce inflammation and protect from continuing damage due to viral antigens; and
- Help re-establish brain chemical balance.

This knowledge can help parents evaluate any cures and therapies for their effects on some of these possible recovery pathways.

For example, the damaged blood brain barrier also explains the sensitivity of kids on the autism spectrum to environmental contaminants and the many arguments about environmental factors being the cause of autism. Kids with TBI show incremental improvements when treated with natural or medical chelating agents designed to remove certain environmental toxins, although there are no major improvements in cases where there is no significant exposure. Thus, treatments directed purely towards removal of such toxins have not been widely successful. But it does make sense to keep individuals with TBI away from environmental toxins given their sensitivity to such toxins.

While there are no treatments that can address regrowth of neurons that could help with full recovery at the time of this writing, there are still things parents can do to help the natural recovery processes of the body and brain in cases of traumatic and hypoxic damage to the brain. To evaluate the best treatments and help currently available out there, it is useful to categorize these treatments based on their purported overall effect on the damaged brain.

But before categorizing them it's important to note that early intervention is known to be an absolute necessity when it comes to seeing improvements in developmentally delayed children. Significant brain changes happen very early in childhood with development plateauing by about age 25. This is true even in normal neurological development, so for parents of challenged children it's a race against the clock if they want their children to recover from their developmental issues. The following graph by the Harvard University Center on the Developing Child illustrates this point.

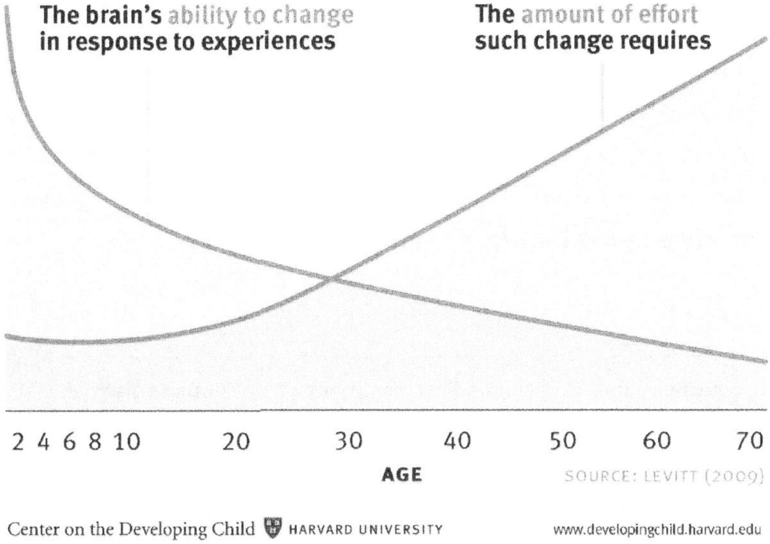

Fig. 11.2: The brain's ability to change/learn with age.

As you can see in Figure 11.2, the ability of the brain to change in response to external stimuli diminishes drastically with age and beyond the mid to late 20's, effecting any change would require an enormous amount of effort and even then, very little potential for change actually exists beyond that point in time. Unfortunately, most parents catch autism in their children very late (typically past age 3). Any progress would also depend on where a child starts and how much brain damage, he or she has sustained and which specific pathways might be affected.

Currently an MRI scan is not ordered other than in cases of obvious brain damage as in cerebral palsy. But recent studies in the US and UK have shown that a 15-minute MRI scan immediately after birth can predict future development delays to a 98% accuracy. So, modifying the standard of care to include an MRI scan at least in cases where the Apgar score is less than 10 would go a long way in identifying brain damage early. In recent times some studies report that cooling therapies have been shown to hold promise for brain damage[11.1]. The infant is subjected to a reduced temperature of a few degrees by placing him or her on a mat that reduces their body temperature by a few degrees. The resulting hypothermia makes the body focus on core functions only, thus helping recovery from brain damage.

Based on the fact that we are dealing with brain damage it is obvious that treatment options need to be addressing the root causes for them to actually work. That narrows down the field considerably in terms of treatments that help and ones that don't. The ones that might work can be categorized into the following three categories as summarized in table 11.1.

Table 11.1

Neuroplasticity based treatments	Biomedical treatments	Future treatments - neuron re-generation
These treatments are based on the fact that the undamaged parts of the brain are capable of learning new skills with training and the brain can repair itself. * Applied Behavior Analysis * Occupational Therapy * Speech Therapy * Computerized brain games. * Neurofeedback * Auditory Integration Training	Treatments based on reducing inflammation in the brain from allergies, viruses, antibodies or altering the brain chemical environment using supplements, hormones. neurotransmitters etc * Antiviral medications * Neuro medications * Gluten Casein free diets * DAN allergy treatments * Stem cell treatments * HBOT (sometimes)	These are currently under study and no treatments are currently available capable of this yet, underscoring the importance of fast tracking of such research. * Tissue Nanotransfection * In vivo re-programming * Neuronal stem cells * Cure neuronal viral latencies

We will discuss some of these treatments in the upcoming chapters.

Key Take Away: Only treatments capable of producing brain damage recovery effects can actually be useful in helping autism symptoms. Parents need to stay away from any treatments that profess to cure the condition without credible pathways to helping the damaged brain perform better. A framework for credible pathways to brain damage recovery is established in this chapter.

CHAPTER 12

Neuroplasticity Based Treatments.

This ability of the brain to change in response to external experiences is called neuroplasticity. Early interventions are mostly of the neuroplasticity type, but if they go hand in hand with biomedical and food interventions (discussed in the next chapter), the results are typically better. There is a need to approach the problem from various angles as there is no single easy solution to the problem. Instead, every therapy helps with a certain aspect of deficit and parents can only hope that there is some synergy between the various treatments resulting in reasonable overall improvement.

Neuroplasticity therapies are aimed at helping the brain establish new pathways to replace damaged ones. They involve using neuroplasticity to learn skills that have been lost or never developed due to damage to grey matter or broken white matter pathways resulting in lack of communication among grey matter areas.

The following are treatments and outcomes that I have personally worked on, or know first-hand of parents that have tried these. Efficacy of these treatments usually depends on whether an individual is in a position to benefit from these or not, and is often a matter of where someone starts. Even in my own case, with two similarly damaged boys, treatments helping one of them don't seem to be helping the other and vice versa due to the difference in the extent of the damage.

Parents are advised to use their best judgment before spending money on any of these. It's hard to restrain yourself from going above and beyond to try and do the best for your child, but it's important to not go overboard since it might only bring disappointment if you do, as individual results vary and none of these is a complete aid to recovery from the condition. Prior knowledge of these treatments will help parents anticipate possible outcomes and goals for their particular child and situation.

Applied Behavior Analysis (ABA) therapy

If you are an autism parent, you likely deal with ABA therapists more than any other personnel. Applied behavior analysis is defined as "the process of systematically applying interventions based upon the principles of learning theory to improve socially significant behaviors to a meaningful degree, and to demonstrate that the interventions employed are responsible for the improvement in behavior" (Baer, Wolf & Risley, 1968; Sulzer-Azaroff & Mayer, 1991). This systematic approach does help in bringing some behavioral improvements (mostly in the form of compliance), which is an important characteristic missing from kids on the spectrum. This is definitely the first step to helping autism spectrum kids and so I list this first.

The success of ABA therapy to a large extent depends on the capabilities of the therapist to identify the exact deficits in the child and apply the principles of ABA to address them. It takes exceptional commitment on the part of the therapist and a reasonable functioning level of the child, for ABA to work. But ABA therapy can only go so far. Like everything here, it is not really a cure but just one cog in the complex array of tools parents should consider to help their child's damaged brain learn and establish new neuronal pathways of learning.

Many ABA centers are good at bringing about considerable improvements in children on the autism spectrum, and strive to continually build on newly achieved skills. However, a recent phenomenon has been that ABA therapy centers have now become a popular business growing alongside the increase of autism particularly in the US. While it can be a good thing and add to the choices for impacted parents, some of these are turning out to be rip-off programs that add no additional value than traditional daycare but come with a hefty price tag. Insurance coverage of ABA due to hard fought victories by parents that have gone before has meant growth of these ABA centers whose main specialization seems to be their ability to efficiently bill insurance to the fullest rather than their ability to help their clients. Parents need to be watchful of these types of centers and choose these settings carefully based on their track record and monitor their kids progress very carefully.

Occupational Therapy

Occupational therapy aimed at developing the motor skills of a child can bring on great improvements in the brain of a child on the autism spectrum. Motor coordination and movement tends to be a feedback loop to the brain, which when given such movement inputs tends to learn from them. While it happens automatically in children, a child on the spectrum needs this kind of therapy to help them get out of the shortcuts and other workarounds their brains tend to create to bypass having to use the damaged parts of their brains. Engaging in any kind of physical sports activity can have the most therapeutic value for kids with ASD. With occupational therapy the therapist evaluates various ranges of motion, from gross to fine motor skills, and engages in therapy directed at improving the motor skill deficits found.

A child's normal motor development needs to go from head to toe, from shoulder out to the fingers. They are first able to hold their head up on their shoulders, flip around on their belly, start using their shoulders to move their body as the first step to crawling, then use their legs to start crawling. With kids suffering from brain damage, the first signs (outside of low Apgar scores after birth) is the failure to develop motor skills out of developmental order. These little nuances of motor development (if caught) can be indicative of other problems down the road.

It should be noted that a damaged brain also makes these kids susceptible to brain hemorrhage risks later in life as adults, particularly if they develop other conditions like high blood pressure and depression. This can be the cause of sudden death, causing even more pain to loved ones. Although children on the spectrum will not typically suggest it themselves, it is a good idea for parents to cultivate interest in their kid in a pastime that involves physical activity early on so they can maintain good physical health, which in turn keeps their mental health from worsening as the years go by. It might just be a simple habit they can do even by themselves like walking or running on a treadmill while watching something on TV or playing a simple solo game like shooting basketball.

Learning therapy – Visualizing and verbalizing

The Lindamood-Bell system originated by Charles and Pat Lindamood and Nanci Bell has been helping kids with learning difficulties in reading and comprehending for over 3 decades. While it is not specifically oriented toward kids on the autism spectrum, sensory-cognitive instruction has proven successful for individuals with learning challenges, including ADHD and autism, and high functioning kids on the spectrum can benefit from this type of learning. Their instruction is based on a theory of cognition. There is a conscious effort to think in pictures, which is apparently lacking in kids with autism, thanks to damaged or detached areas of the brain needed for imagination. It should be noted that this kind of therapy can also be expensive for families since learning is typically not covered by insurance and quality one on one therapy such as this tends to be pricey, putting it out of reach for many families.

Computerized Brain Training Games

An individual's ability to learn comes from their ability to observe, focus and discern visual and auditory inputs to the highest level of detail, use that for pattern recognition and incorporate that understanding into existing knowledge. When it comes to individuals with autism one or more of these functions is lacking (or is working sub-optimally) due to brain damage. It is quite possible that brain training games can be designed to address some of these deficits.

The companies that have created computerized brain improvement games claim to improve IQ and help take even the cognition of non-autistic children "to the next level." However, they have recently come under fire for their claims of improving IQ in all customers. Following customer complaints even the Federal Trade Commission has come down on their claims pretty harshly. Researchers seem to disagree on the efficacy of such programs. But it's easy to see why it may not work very well in the general population. These programs need to be tailored to address specific deficits in individuals like auditory memory, audiovisual sensory integration involving dedicated continuous practice over long periods of time.

Neurofeedback Therapy

Biofeedback therapy is a technique used by therapists to train people to improve their health by controlling certain bodily processes that normally happen involuntarily, such as heart rate, blood pressure, muscle tension, and skin temperature. An individual's response to certain external stimuli is measured with a measuring device and the individual is trained to respond differently in an attempt to improve health. Electrode patches are placed on different parts of the body to measure nerve or muscle response, heart rate, blood pressure or other function. The electrodes are connected to a monitor that displays the measures. With help from a biofeedback therapist, patients are trained to consciously control their responses, thereby helping them practice certain types of responses over others.

Neurofeedback is a type of biofeedback where brain waves are used to produce a measurable signal that is used as feedback to teach self-regulation of brain function. This is useful in autism and ADHD therapy where the subject is relatively verbal and understands the need to work with the equipment and train him or herself to self-regulate brain response. Electrodes are glued on to target areas of the scalp and the other end connected to measuring equipment and a monitor displaying brain activity.

There can be variations on how brain training can be achieved. For example, the subject may be asked to watch a movie or video and based on the brain wave activity the monitor is programmed to dim down or the volume drops in response to types of brain activity that reflect a lack of focus. This becomes a reward system with focus and concentration being rewarded with good picture quality and sound. Alternatively, a game could be played where the subject controls an object on the screen using his brain waves.

It should be pointed out that neurofeedback is not a cure for the underlying condition but an effort to train the brain to overcome its shortcomings. The outcomes depend on the techniques of the therapist, the patient and the equipment used, the ability of the therapist to filter noise from real brainwave activity, etc. While some parents report incremental positive changes for kids on the spectrum, this might typically help higher functioning kids on the spectrum or perhaps typically developing kids with minor shortcomings.

Key Take Away: Neuroplasticity or the ability of the brain to learn new skills remains a very important aspect of treatment for spectrum kids. However, the level of benefit depends on the extent of damage a kid starts with and the expertise of the personnel handling the treatment. The ability of the brain to learn slows down with age and early intervention of expert therapists can make a difference, although outcomes are not guaranteed as learning is only one aspect of the overall brain damage recovery picture.

CHAPTER 13

Biomedically Based Treatments

These are treatments directed toward creating a brain environment conducive to repair and recovery. It could be antiviral medication to reverse vaccine and other viral effects, neurotransmitter-based medication for calming the brain or supplements to help the gut and brain. Based on the severity of brain damage, biomedical interventions are marginally effective but have been far from the cure they are sometimes touted to be. It is easier to cure a diseased organ than to fix an organ that is damaged permanently. Still, armed with knowledge of what really is causing the issue you could sort through the mechanisms of how a certain cure might work and make an informed decision about pursuing it. Most of these treatments I go over in this chapter do not help cure autism altogether, but they help alleviate certain aspects of behavior or help verbalization or help with better coordination or any one of the many brain functions that are impacted by autism.

The DAN Protocol

Dr. Bernard Rimland (1928-2006) was an American psychologist and researcher whose son Mark was diagnosed with autism. He set out to correct a lot of myths surrounding autism in the 1960s and led the movement that dispelled the now preposterous refrigerator mother theory. He wrote a book on Infantile Autism and founded two organizations: the Autism Research Institute (ARI) and what is now the Autism Society of America (ASA). He has had the single biggest influence in autism research, directing research toward both neurological and biomedical interventions.

The DAN protocol was originally a program from the Autism Research Institute. It is based on the belief that autism is a disorder caused by a combination of a lowered immune response, external toxins from vaccines and other sources, and problems caused by certain foods. The Autism Research Institute has discontinued the DAN protocol since 2011 (apparently due to their inability to control individual practitioners and their quality of service).

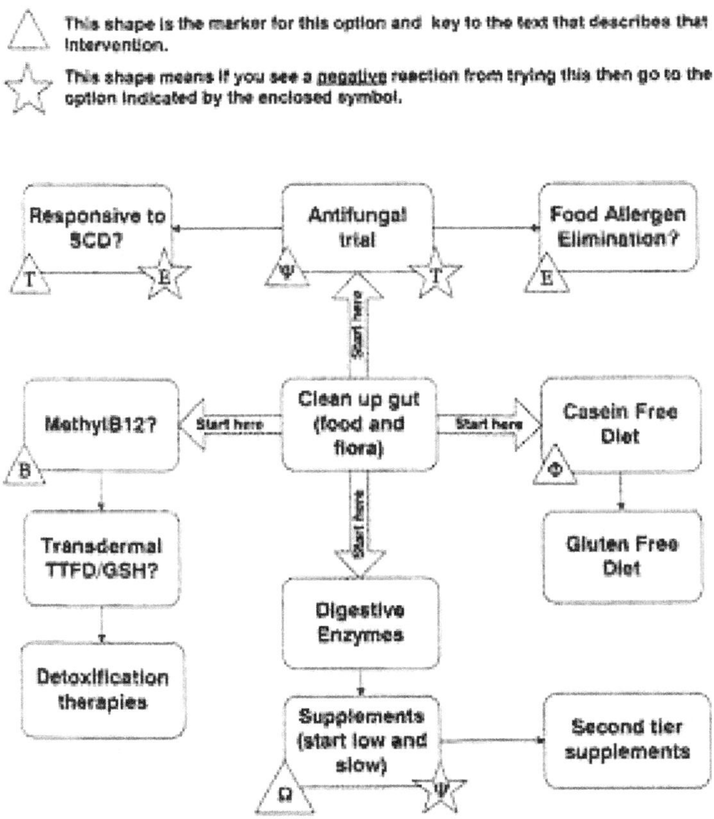

Fig. 13.1: The discontinued DAN protocol.

A summary list of treatments under the DAN protocol include Actos, amino acids, anti-inflammatories for inflammatory bowel disease (Rx), anti-yeast diets, antifungal pharmaceuticals, antiviral medications, chelation, chelation (oral), chelation (rectal suppositories), chelation (transdermal products), colostrum, diet avoiding food allergens, digestive enzymes, essential fatty acids supplementation, Feingold diet, glutathione, gluten and casein free diet, heavy metal detoxification, hyperbaric oxygen therapy (hard chamber), hyperbaric oxygen therapy (soft chamber), intravenous immunoglobulins (IVIG), low dose naltrexone, low oxalate diet, methylcobalamin (B12 injection), nutraceuticals, probiotics, secretin, specific carbohydrate diet, spironolactone, transfer factor, vitamin/mineral supplementation, and vitamin C and/or minerals (IV).

While many of these treatments offered under the DAN program have been incrementally useful in the vast majority of autism impacted kids, none of these seem to help completely recover from the condition in most cases. Some aspects of the DAN approach remain valid, particularly the need to rule out the low hanging fruit like Lyme disease, PANDAS, PANS and food allergies that produce autism symptoms in a small fraction of kids that receive the autism diagnosis. In my experience and the experience of most parents I have talked to, the DAN approach can be summarized as follows.

Effective treatments in the DAN approach include:

- Detection and treatment of Lyme, PANDAS, PANS.
- Antiviral drugs (acyclovir, valacyclovir)
- Gluten free diets
- Food allergy identification and avoidance.
- Yeast(candida) overgrowth reduction (low sugar diets, antifungal treatment)
- Supplement deficiencies in Vitamins, Minerals, Essential fatty acids, Amino acids, Enzymes and co-enzymes.

In summary, treatments of the DAN approach targeting immune and viral effects and gut dysbiosis prove to be their most effective approaches while some of the other treatments like heavy metal detection and chelation (with no obvious exposure), GcMAF, Secretin etc., are often either ineffective or very subjective where clear patterns cannot be established.

Antiviral and Neuro Medications

Viruses known and unknown can wreak havoc in the brain, particularly a damaged one, with a leaky barrier, both by their direct impact and by the impact of immune responses including inflammation. Even viruses that exist in almost 100% of the human population which don't affect individuals with normally functioning brains can introduce detrimental effects in someone with brain damage. Your doctor might be able to order antibody testing to identify any viral excesses in your child's body. When treated with the appropriate antiviral medication the outcome tends to have an incrementally positive effect in reducing behaviors and helping the development of a child on the autism spectrum.

Neurotransmitter-based drugs that are known to calm down individuals might prove effective in controlling and calming down autistic children with behavioral, anxiety, sleep, OCD and even seizure issues. Working with a trusted Defeat Autism Now (DAN) doctor or a neurologist would be the best course of action in this case to narrow down the drug that might help best. Ideally, it is advisable to avoid drugs where possible, but in the worst of cases, where the benefits outweigh the risks, they might be an important ingredient that could be of incremental help in bringing about positive outcomes.

Food Restrictions

When done right food can truly be medicine, and food restrictions help kids on the autism spectrum to the extent that brain healthy foods tend to improve focus and awareness in everybody. The vital role of food in maintaining mental health is finally being recognized by science and has led to the burgeoning new field of nutritional psychiatry.

While not a miracle remedy, in many cases avoidance or inclusion of certain foods can work like medicine and such effects are particularly pronounced in kids on the autism spectrum. These kids have a lot of digestive issues and food allergies, thanks to viruses latent and hiding in their damaged brains and enteric nervous system.

Gluten Free / Casein Free Diets: The most common among these diet restrictions is the gluten free casein free diet (GFCF) diet. Glutamate is a neurotransmitter in the brain that is excitatory. Foods with gluten content are seen to have a significant excitatory effect for many kids on the ASD. Gluten is a protein commonly found in many protein rich foods including meats, dairy, wheat, rye, barley, soy, couscous, semolina and many others. Like all other types of protein, the digestion process breaks down the gluten protein into amino acids in the body. One such amino acid resulting from protein breakdown is called glutamine and glutamine in turn can switch to becoming glutamate in the brain (in what is called the glutamine – glutamate cycle). Excessive glutamate in the brain results in excitatory overload and hyperactive behavior.

Of all gluten containing foods, wheat, soy and dairy products seem to contain the most glutamine content and so it makes sense for ASD kids to avoid them. Unfortunately, in our world of processed foods, gluten proteins are used as the glue that gives the consistency and chewiness to most processed foods, making avoiding gluten a difficult task. But the results of avoiding it can be worth it in many cases, particularly where there is a gluten allergy involved alongside the ASD and hyper-activeness.

Food allergies avoidance: Children on the autism spectrum in many cases are not verbal and so do not complain verbally about any discomfort to parents. Sometimes it might look like they are crying for no reason, but it is likely due to some discomfort that they can't verbalize. It's up to parents to play sleuth to discover what the underlying reasons could be. Hence the IgG, IgE and skin prick allergy testing panels tend to be important tools in the hands of parents trying to get at possible allergies in their kids (particularly all the gut allergies that kids on the spectrum typically suffer from) that might be causing them pain. The IgG is a particularly convenient panel test that has become popular, although it often can come up with some false positives, particularly for foods consumed regularly and frequently. But it also indicates sensitivity to other rarely eaten foods as well which tend to be correct. It gives parents a good starting list of foods to watch out for. Removal of these food allergens can help symptoms up to a certain extent. These are efforts aimed at reversing the harmful effects of processed foods, food allergies and fighting the impacts of residual viruses from the live vaccine doses hiding in the brain and gut neurons of autistic kids.

Low Carb Diets: High amounts of sugar in the diet can be the enemy in hyperactive kids. It can introduce silly behaviors like uncontrolled giggling and stimming while worsening restlessness and causing hyperactivity (among other disruptive behaviors). These effects might be due to intestinal dysbiosis of sugars where they end up promoting candidiasis and other undesirable gut biome issues.

High carb diets can generally be more inflammatory than lower carb diets. Such inflammation can in turn be part of the problem causing the adverse effects on kids with ASD. Of the various diets out there for that show promise for various health conditions, the Keto diet has become popular among autism parents after a recent study[13.2] at the University of Honolulu, Hawaii seems to have found significant improvement in about half of the 15 participants that were on a keto diet after 3 months.

The keto diet involves restricting diet to 70% fats, 25% proteins and 5% carbohydrates. This switches the body to using the ketogenic digestive pathway that is designed to use fats as the main source of fuel for body processes. It is a difficult diet to follow as it involves removal of processed foods altogether.

Simple Carbohydrate Diets: The simple carbohydrate diet is an elimination diet prescribed in the DAN protocol. It works for kids on the autism spectrum and others with very significant gut allergy issues. It proposes limiting diets to simple carbohydrates containing nothing more than a single glucose molecule. Typical complex carbohydrates contain polysaccharides or multiple glucose molecules which are considered harder to digest and absorb for kids with Crohn's Disease or Irritable Bowel Syndrome with or without autism symptoms. It is indeed helpful for a certain segment of those impacted with autism symptoms, but it can be a hard diet to follow particularly in kids.

Stem Cell Treatments

Stem cells are precursors of all living cells in mammals. After the initial embryonic stage where the entire embryo is made of stem cells, these differentiate into other cells creating the entire body composed of hundreds of cell types. However, stem cells continue to exist in various parts of the body throughout life. These cells have the ability to divide symmetrically into 2 stem cells or asymmetrically into one stem cell and another body cell. Thus, they have properties of self-renewal and multipotency.

Use of these cells in treating various diseases has been a subject of controversy particularly around the use of stem cells from human embryos. But not all research involves embryos. It is possible to harvest stem cells from other body parts like the bone marrow and grow them indefinitely in laboratory settings. Stem cell treatments are not an FDA approved option in the US for the treatment of autism, although they seem to have some marginal positive effects in many kids. There are concerns regarding safety, efficacy and standardization that need to be addressed. However, there have been instances where there has been miraculous recovery, where a kid goes from marginally verbal to fully verbal after stem cell treatment. For reasons already explained, the younger the child (particularly children under 6) the better the probability of seeing some positive changes. The useful properties of stem cells in these cases are thought to be their immune modulatory effects rather than the potential to differentiate into neuron cells.

As of early 2020 two types of stem cell therapies are available to patients for autism treatments based on where the stem cells are sourced from. Autologous stem cell treatment is where stem cells are used that are harvested from the bone marrow of the hip bones of the patient. The stem cells are multiplied and re-injected into the patient a few hours later. The challenge with this treatment is the torture you have to put the child through to extract bone marrow and deal with the after effects of rejection of cells by the immune system, which happens most of the time.

Another type of treatment involves use of stem cells sourced from umbilical cord derived stem cells from own preserved cord or donors. These cells are then multiplied in a lab. This only needs a limited amount of donor cells that are then re-used to create more stem cells in the lab and injected into the patient. These cord stem cells are more primitive than stem cells derived from bone marrow. The treatment typically involves four injections of stem cells over a period of four days. While considerably less painful, this process has the same probability of success as the autologous route.

Both these current processes owe their efficacy (if any) to their immune modulation effect, which in turn can result in reduced inflammation in the brain. It might work in cases where constant brain inflammation is the cause of the autism symptoms. But the ability of these stem cells to differentiate into neurons and grow in the brain is very low or unknown and as a result the efficacy remains low. What might really work is if the stem cells can be directed toward becoming neurons for repairing the damaged brain. Such a possibility exists with neural stem cells (NSC). Neural stem cells are specialized stem cells that generate radial glial cells that give rise to the neurons and to the glia of the central nervous system. They are under research for their ability to help recovery from stroke and ischemic damage. While these could most certainly help with autism, they are not commercially available to patients due to challenges with sourcing neuronal stem cells. These are found in localized brain regions called "niches" such as the sub granular zone of the hippocampal dentate gyrus, the subventricular zone of the lateral ventricles, and the external germinal layer of the cerebellum. Extracting and growing these cells in laboratory settings remain a challenge. The efficacy of any such method in treating autism particularly from traumatic injury and ischemic injury needs to be evaluated. But this kind of treatments hold out a lot of hope for kids and adults in the autism spectrum in the years to come.

HBOT - Oxygen Therapy

Fiscus religiosa (commonly called the sacred fig or the Peepal tree) is a large dry season-deciduous or semi-evergreen tree that grows up to 30m tall and is native to the tropical and semi-tropical climates of India and Southeast Asia. In the eastern traditions this tree is considered very sacred. Gautama Buddha, the originator of the Buddhist religion, is said to have attained enlightenment while meditating underneath a Ficus religiosa. The leaves of this tree move continuously even when the air around is still and no perceptible wind is blowing. Walking around this tree in the morning and evening has traditionally been suggested as a remedy for mental retardation. The modern explanation to this is that this tree gives out copious amounts of oxygen which helps the damaged brain heal and work better.

In a medical context hyperbaric oxygen chambers therefore could serve as a remedy for traumatic brain injury. There are known cases of people with traumatic brain injury doing better with HBOT therapy. However, HBOT therapy has mostly fallen short when it comes to treatment of kids on the autism spectrum. Only very slight improvements (if any) are seen even in kids that are treated on a regular basis. This might be due to the fact that this treatment may only work in cases where there is currently existing, festering brain damage and inflammation from it. Unfortunately, those with damaged but healed and scarred pathways in the brain (like seen in many spectrum children) may see limited to no improvement from this therapy.

Key Take Away: Biomedical interventions can help promote better bodily functions which can help improve brain functioning. The DAN movement and treatment protocol was among the first efforts towards identification of true causative factors of autism. Elimination of inflammatory processes and promoting of healing mechanisms are the main hallmarks of biomedical interventions and can bring incremental relief to many kids on the spectrum.

CHAPTER 14

Potential Future Treatments

Up and Coming Technologies

For parents impacted by autism in their kids, current medical science has no options to treat the condition. The issue is further compounded by the fact that autism researchers have bought into the story of genetic causes of autism. They have been hard at work trying to find genetic causes for the malaise, looking for evidence that doesn't exist, wasting valuable time and resources that could have been deployed to look at brain damage recovery instead. Now that we understand that autism is caused almost entirely by traumatic or hypoxic brain injury, happening mostly during birth, it stands to reason that parents can look towards current research in the field of traumatic brain injury recovery for answers. And looking at the symptoms and directions of research for traumatic brain injury (TBI) provides even more evidence for the fact that autism is caused by traumatic brain injury. The latest research in TBI recovery[14.1] seems to be focused on stem cell treatments and use of hypothermia, while also investigating mitochondrial dysfunction, gut dysbiosis, microglial activation, cerebral microcirculation impairment that seem to follow TBI exactly like it happens in autism. Mitochondrial dysfunction and gut dysbiosis might follow TBI, directly as a result of a compromise in the blood brain barrier by injury even in adults. I have previously mentioned in chapter 9 that mitochondrial dysfunction could be a result of viral latencies when certain viruses get past the blood brain barrier and establish a latency in the neuronal cells of the brain.

Neuron cells in the central nervous system (CNS), typically do not regenerate or regenerate enough once they are damaged. Decades of research into the possibility of regeneration of neurons has turned out to be fruitless this far. Current research into brain damage, particularly Chronic Traumatic Encephalopathy (CTE), Stroke, Multiple Sclerosis, Alzheimer's and Parkinson's, might provide useful information leading to the next breakthrough in autism treatment. Any treatments that can address neuron regeneration, scar tissue removal, reversal of neuronal viral latencies can be of particular help for autism. The open sourcing of gene-editing with CRISPR and the evolution of bio-hacking might even hold out hope to speed up research in this area, not only for genetic diseases causing autism symptoms but also for brain healing from trauma by methods unknown to us at this time. Some of the current research showing promise are discussed below:

Neuron regeneration: Any treatments that have the potential to add additional neuron supply to the brain can be of potential benefit to autism patients. Neuronal stem cells (NSC) is a technology currently under study for brain damage in stroke patients. NSC's are capable of differentiating into neurons, astrocytes, and oligodendrocytes, and adding an additional dose of these cells into the brain might help further natural healing. But immune rejection and long-term functional integration of such cells into the brain continue to be challenges to overcome Such technology might offer help to autism patients in future if successful.

In similar fashion, another possible solution is tissue nano transfection (TNT), where skin cells are teased into growing other cell types capable of repairing damaged organs. TNT involves use of a nanochip to deliver a biologic payload to skin cells in the target area, reprograming these cells to create another targeted type of body cell. This kind of technology is currently successful in lab animals. If successfully applied to the human brain, such a technology would be a boon for numerous kids and adults on the autism spectrum.

In vivo reprogramming: This is an emerging strategy for repairing hard to heal organs like the brain that consists of converting resident tissue-specific cells into the cell types that have been lost due to disease or damage by a process called in vivo lineage reprogramming. Even in areas of damage like in a transected spinal cord, the bridges of healing tissue that develop seem to contain radial glial cells[14.7]. If glial cells in the brain can be repurposed to become neurons then they could be used to help the CNS heal from damage. A payload of retroviral protein capable of such reprogramming when ingested could accomplish this feat. It has been demonstrated in lab animals, and if successfully reproduced in humans it could hold promise of a cure for brain damage.

Scar dissolution and removal: In the case of trauma scarring is necessary to quickly isolate dead tissue, re-establish the blood brain barrier and repair broken capillaries. But it is widely believed within the scientific community that the scar tissue formed in response to brain and spinal injury inhibits regrowth of neurons. Some new technologies such as using rapid acoustic pulses to reduce scar volume have been promising. If such types of technology could be used to reduce scarring in the brain and spinal injuries they might help with healing and recovery from injuries of the central nervous system. (Recently there have been some studies that seem to suggest that scarring actually helps regrowth, but it is hard to give credence to them knowing that scarring represents both a physical and chemical barrier to regrowth.)

Reversing of Neuronal viral latencies: As mentioned before, viral latency in the body and its physiological effects can be a huge unknown. A latent virus can impact cellular processes. When the latency is inside of neuron cells, the effects can be even more profound due to the possible impacts on complex brain processes. During a latency, the viral genome is maintained as a separate genetic entity (an episome) within the neuronal cell nucleus. It can then re-activate when the conditions are right or create ongoing chronic conditions, which can be the effects of the virus itself or the antibody response. The Herpes Simplex Virus (HSV-1) is one such virus that is believed to establish a neuronal latency. Other viruses capable to infecting neuron cells and establishing a latency include the measles and polio viruses. These viruses take advantage of the differentiated functioning of brain immunity to impair the body's self-healing abilities to recover from trauma. Hence, removal of viral latency in neurons (if possible) represents one of the best ways to help the body heal itself from damage. In fact of all the research into traumatic brain injury recovery, this viral latency removal might represent the lowest hanging fruit. Not to say it is easy by any means, but with current technologies like CRISPR protein and gene-editing have put this type of treatment in the realm of possibility.

Advances in brain research and treatments for healing injury have been gaining traction in recent times. For autism parents these offer a ray of hope. Given the right directions of research and funding soon there might come a time when brain damage will become treatable and completely reversible.

Key Take Away: The similarity of symptoms between autism and traumatic brain injury in adults further underscores the fact that autism is indeed traumatic brain injury. Gut dysbiosis and mitochondrial dysfunction are consequences of damage to the blood brain barrier. Treatments that can reverse these after effects might be critical in finding permanent recovery from TBI. Numerous mechanisms capable of regenerating damaged neurons and healing brain damage are currently under investigation. Some of these have shown promise in animal studies, and could become viable options for autism treatment in the future. Drugs currently under research for brain injury recovery, Alzheimer's, Multiple sclerosis etc., might also hold out hope for kids and adults on the spectrum.

Part V
The Legal Recourse

Malpractice by a medical practitioner is defined as a deviation from the accepted standard of care, particularly if that deviation resulted in injury. Pitocin abuse by prolonged induction of labor beyond contra-indication is a perfect poster child for a medical malpractice lawsuit. Yet the defeat of autism cases in vaccine court in 2010-11 and the court pointing at genetic causes poured cold water on lawyers' willingness to undertake any 'autism' cases. Lawyers who might be willing to take on birth injury cases balk at the possibility once the word autism comes into the picture. Turning that attitude around is the goal of this final part.

CHAPTER 15

The Impact on the Family

Many autism parents have written media articles and some even entire books detailing the challenges parents face and live through as a result of their kids' autism. Some even describe how they have been suicidal at times, even contemplating taking the lives of their kids along with themselves. While such extremes might be construed as weakness and desperation, I can totally understand how those kinds of thoughts might come about. Watching a 10 year-old child scream and throw tantrums, unable to verbalize the reason for their outburst and explain the pain they're going through can bring on those feelings of helplessness and desperation to the extreme. It can easily make someone hate their lives and pity the existence of their suffering child. It's possible to go through suffering and stay strong if you know there's light at the end of the tunnel. Unfortunately, for parents of kids on the autism spectrum (particularly lower functioning kids), there's not much to hope for either.

Sometimes ignorance is bliss. From the moment we found out what had actually happened to our cute little kids (thanks to the MRI scan of my second son) things have never been the same again. In our case our pain multiplied by a factor of 10 when we found out that the condition of our kids was not natural but the result of medical malpractice at birth. Previously, my wife and I would jokingly blame each other's cranky moments as the reason why the kids had turned out the way they did. She blamed my lineage and I would tease her about her family history, but neither had any real clue. Life went on, challenging and frustrating, but without the nagging sense of having been wronged for life by the doctor we had counted on so much.

Stalled Development

Initially we weren't surprised when we heard the autism diagnosis, thanks to prior research we had done on our boys' hyperactive behavior, but weren't sure of what to expect. We were not expecting this was going to turn our whole lives upside down. But that is exactly what happened and we were caught totally unprepared. The impacts have been felt not just by us, but by the extended family, relatives and close friends as well.

At first there is the denial of any deep issues with your child. I simply couldn't believe anything would go wrong with our kids. Unfortunately for most parents, this state of denial depending on how long it lasts, results in the loss of valuable time in the life of the damaged child. Like in our case, it takes a while for parents to come around to the realization that something is wrong with their child and they need to act quickly to help their child. For me it has been particularly difficult after I learned about all the disinformation being shared out there on the internet, effectively ensuring other parents stay in the dark about the root causes of their children's misery.

Outwardly, there was truly nothing wrong with our kids when they were infants and toddlers. But we came to realize later the brunt of these impacts are truly all-round and we had to adjust our lives and priorities around the kids. Most of the ill-effects become more and more obvious with time as the kids grew up. Unfortunately, autism is a problem which tends to only get worse with time. As our kids grew up, the divergence between them and the rest of their peers only got more and more manifested. Regular pediatricians are not equipped to deal with such cases. They address the problem in the typical symptomatic shotgun approach of modern allopathic medicine, and all they are able to come up with is only speech therapy and Applied Behavior Analysis (ABA). Our experience in this regard has been very typical of most parents out there suffering the consequences of induced labor brain damage.

The brain controls all human behavior and all the complex higher-order functions, including speech, verbal ability, cognitive processing and the complex social interactions that we take for granted as humans. When the brain is damaged, the first impact is to these higher cognitive functions which typically only develop outside the womb as a baby grows up in the first few years of life. The gap in development between undamaged kids and those with brain damage begins to show between the ages 2 and 3. That is when kids start taking off with their verbal ability and social interactions, while kids with damaged brains tend to be in their own world. Progress from that point becomes an uphill battle for the entire family as this gap only gets worse with age if left unattended and hard to fix even with a lot of effort.

After about 6 months old our older boy became a tough baby to carry and hold onto. He was always squirming and wiggling, trying to jump out of my hands with no fear of falling down. Once he started walking, he always ran to get from one place to another, sometimes unable to maneuver around obstacles and bumping into them. I took it to be precocious development and was generally happy he was so fast. However, hidden beneath this veil of apparently extreme motor skills were some really serious deficiencies. My son was always throwing up a lot. He would refuse to stand in place long enough to learn words to talk. He refused to be potty trained, and stayed in diapers until he was about 4.

Our 2nd boy was a little slower and only learned to stand up and walk with support at 11 months old. He didn't become proficient at walking until he was about three years old. Just before age 3 he was beginning to show signs of talking and becoming a typical kid. But his third-year vaccination did him in and he retrogressed into losing the interest and ability to talk.

While both boys did learn to walk and run on their own, there were some serious deficiencies in how our boys progressed. Both lacked the ability to pay attention to my directions. I could never train either of them to stand still long enough or listen long enough to play any games. The lack of response to simple yes/no questions even at age 3 seemed to get in the way of learning anything. They lacked the motor coordination, focus and strength in their hands and fingers to hold onto simple objects like pencils, preventing them from even drawing in a meaningful way. Then there was the repetitive behaviors like looking at the wheels of toy cars and spinning them all the time. This became so bad that we had to take away all toy cars.

Most people, even those with high functioning kids on the spectrum, cannot quite understand what life is like on the lower end of the spectrum. An easier way to imagine what it is like, is to think of a 10-year-old that behaves like 2- or 3-year-old. The lack of fine motor control at various levels makes even simple functions like blowing the nose, rinsing the mouth or swallowing a tablet quite a challenge. Doctors' visits and injections can be a nightmare, with them twisting, turning, and screaming, making it practically impossible to get a needle in at the right spot even with 2 or 3 attendants holding them down. Having to put them through medication and injections, knowing that their condition was caused by medical malpractice only adds to the trauma of the parents.

The only redeeming factor is they retain their innocence and cuteness for much longer. But for the most part they simply destroy toys or play with them inappropriately, put objects in their mouths, and have no concept of dirty versus clean. There's constant vocal stimming and hand flapping all day. They are unable to follow stories even in the simplest of pre-school cartoons. Essentially, it's the terrible twos with the weight, strength and reach of an older 10 plus year old with no end in sight. A typical special needs parent can expect weekends to be taken up mending and fixing things that have seen the wrath of a low functioning child.

Kids on the autism spectrum are otherwise regular kids that are lacking certain higher-order functions like speech, memory, etc. That means they come into this world capable of the same feelings and emotions as all others. Unfortunately, their main issue is the ability to express themselves verbally as their brains don't process information at conversational speed. They are full of thoughts, feelings and desires that they would like to express but simply cannot do so with words. With time this inability to express themselves turns into nervousness and social anxiety that is amplified to a much greater degree than would otherwise. The result is they try to shut themselves out from the outside world. Combined with their basic inability to focus on what their eyes see and their ears hear imprisons them in their own mind.

Having difficulty talking is usually due to an inability to pay attention, which in turn to a lack of environmental learning which is critical for any sort of higher-order social development and a key aspect that makes up the complex social interactions of humans. If kids on the spectrum can't copy what the adults or peers do due to their lack of focus, it just means they can't learn anything by themselves or be taught anything very easily. While attention deficiency is rampant even in the regular peer groups of children, such extreme lack of focus is typical in the spectrum. This results in an inability to learn language or express their needs and even share little things like "my tummy doesn't feel well."

The Future of Autistic Kids

Like every other aspect of our lives, even family vacations were impacted and subjected to constraints. The very few vacations we've managed in the past few years we take during odd off-peak times so that we don't run into overcrowded situations in tourist destinations. The sounds and chaos can be overwhelming and our kids don't take well to crowds and don't enjoy all that sensory overload. On one such early November off peak season visit to the theme parks of Orlando, Florida, we came across a large group from a special needs caretaker home sporting 'autism' T-shirts. They had probably picked the same times to visit for the same reasons we picked ours. We saw older otherwise normal looking men in their 40's and 50's and even older having to be helped around the park. They were mostly not talking, just following directions and mostly zoned out, being helped around by caregivers as if they had just turned 3 years old.

Until that moment my wife and I had been living in this bubble that autism kids mostly end up recovering and held on to the hope that our kids would eventually be alright. Every blog and story out there was from parents relating their positive success stories and we never heard of the alternative. Not having had any previous experience with autism in our family circle, we never knew any better. Our boys look normal and we had come to expect they would eventually grow out of their problems. We were suddenly hit with the alternate reality of what would happen if our boys didn't make the recovery we had come to expect.

It's really hard to fathom, but it's true that your birth plays a very crucial role in how your life turns out. The way you came into this world can determine your brain function, which in turn determines a significant amount of how your life turns out. As modern living continues to shift us away from agricultural subsistence-based occupations to more industrial and technological occupations, the reliance on mental faculties for participating in the workforce have become increasingly critical. Brawn-work based occupations are being replaced more and more with brain work. Even schools have over the years shifted focus to improved vocabulary and comprehension as the key ingredients of all their study programs. There tends to be more emphasis on independent study versus the more directed study that has been the norm for many decades.

Unsurprisingly, this puts kids on the autism spectrum at a further disadvantage. With their damaged brains they tend to become adept at doing certain things with narrow areas of focus at best. While such jobs existed in the past, they risk getting left behind more and more as societal expectations change and their condition becomes more and more of a problem. While advances in medical technology might eventually might help these kids, the societal advancements and expectations that accompany such technological changes puts them at an even bigger disadvantage and unable to keep up with the expectations of the world around them.

The future of their children is what worries autism parents the most. What will they do when they get out of school? Will they be able to become self-sufficient? Will they be able to hold on to a job or keep themselves happy, healthy, active and become contributing members of society? Will they be able to have a family of their own? How will they survive in that fast-changing world out there after the parents are gone? Unfortunately, the answers to these questions have not been very promising so far.

While as human beings we are wired to be social beings, we also seem to be wired for ganging up on those of us that fall outside of the norms that we have established for ourselves. The result is these kids can be subjected to a lot of bullying when they don't socialize like others. Worse still, they can have stimming behaviors (giggling, flapping, talking to themselves, etc.) that make them the subject of ridicule and bullying.

Modern society is a social system based on rewarding creativity, logic and social and emotional intelligence. Most successful people succeed because of their intelligence even if a high IQ is not really a pre-requisite to what they do. Higher acumen is amply rewarded in our society, but this is more pronounced in modern times where most opportunities are tuned toward people that excel in a field due to their abilities combined with intellect. This is where kids and adults on the autism spectrum including ADD, ADHD and the like are seemingly doomed for failure. For no fault of theirs, they are destined to not achieve their full potential that could have been had their birth process not damaged their brains.

Lives Changed Forever

For people not used to taking care of a special needs child, it is hard to imagine the mental stress and agony that special needs parents go through. Most people consider autism as something that happens to other people, so they simply don't have a need to understand it. This is perhaps due to the genetic connotation that's commonly associated with a condition like autism. Such an attitude is also what gives rise to the stigma toward autism and other mental health issues in general.

For example, slowly but surely the parents of autistic children (just like their kids) become social outsiders as well. You notice after a while that your friends' kids are not interested in playing with your kids, simply because these kids will not engage in any social play. Soon the parents follow suit and you stop meeting up as much as you used to and your own friends have less and less time for you. Your social circle might shrink down to the closest of friends. And it goes both ways: you don't have time for socializing much either as social situations are taxing on the kids and you don't want to deal with people who don't understand your kids. Soon the main social group becomes a circle of other special needs parents who are the only other people who can comprehend your situation by virtue of being in a similar situation themselves.

Yet special needs parents are a group that has experienced exceptional trauma and suffers from severe time challenges, which leaves very little time for social meetups. Only the strongest parents manage to persevere, continuing to try to keep the kids engaged. Unfortunately, we live in times where the time you have is money you don't have and vice versa. The autism problem is a double-edged sword in that it takes both time and money to even attempt fixing the problem. In spite of the fighters that we are, most autism parents are too busy fighting for our kid's future that we have not had the time to sit back and think about the root cause of our current situation.

Acceptance of kids with autism is at present hit or miss. In popular culture PBS recently launched a character with autism called Julia that appears on Sesame Street, and the Disney movie 'Finding Dory' was about a fish born with memory disorders. But for every story you hear about families that adopt special needs kids there is a story about hate groups seeking to "cure" the world of autism by killing off children with autism. One popular Google search term was "people with autism should be killed" and it was only after a protracted fight by impacted parents and good Samaritans of the world that Google finally removed it.

Your kids being called autistic does give you a different perspective in life. All the little trivial things that people complain about all the time, talking about them as these huge challenges in their lives sound trivial, even stupid to you, when autism is on the back of your mind all the time. In our case (as with many other impacted parents) we've leaned into our faith and spirituality. Being believers in the idea of karmic influences in our lives, we console ourselves assuming what happened to our kids was the agenda of the almighty, perhaps part of a grand plan that we have no idea of at the moment. It gives us the strength we need to get through every day of our lives without breaking down. It is very draining to be autism parents and we figured early on that focusing on positives and planning out things a little more conservatively on what we could accomplish with our kids went a long way to reducing frustration and improving our actual accomplishments. We recognized that we need to be around for these kids for longer and letting the stress get to us was not helping our cause.

However, there is a glimmer of hope from the work of some non-profit organizations at the local level who, by blazing new trails in totally unchartered territory of handling grown kids who have turned into adults with autism, have shifted the attention of many others in this direction. More programs like these are needed to help these unlucky victims of the birth process that become collateral damage in the modern society's fight against pandemics using vaccines.

The one thing parents cannot afford to do is to give up on their child, particularly when the whole world around them wants to move on and give up trying to help. For kids on the spectrum, their parents are their only friends, their only true support system in the fast and busy world around them. This is why a coordinated legal response is an absolute necessity, to help secure the future of these kids and to offer some respite to impacted parents.

Key Take Away: The suffering of autism parents knows no parallel as it is a daily fight with no hope for change and no end in sight, particularly on the lower functioning end of the spectrum. Even great motivational quotes from great people like this one from Winston Churchill – "When you are going through hell, keep going" – don't really capture the kind of hell autism parents go through and bring no solace. The only possible redemption is some kind of medical breakthrough to cure brain damage.

CHAPTER 16

Pitocin Abuse as the Cause of Autism

Induced labor has existed for more than a century in various forms right after the discovery of the uterine effects of the oxytocin hormone in the early part of the 20th century. The modern induced labor process has come a long way in that time. To understand induced labor, it is necessary to have some intermediate knowledge of the uterus, pregnancy and the labor process.

The Uterus

The uterus is an exclusively female organ found in all mammals and a feature that distinguishes mammals from other classes in the animal world. It is a hollow muscular organ located in the female pelvis between the bladder and rectum.

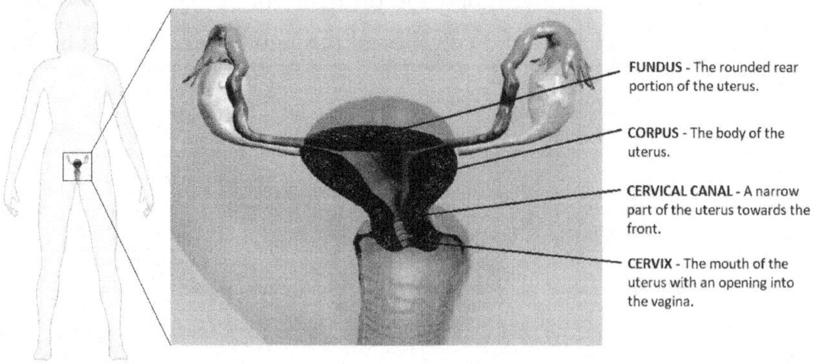

Fig. 16.1: Front cut cross-section view of the uterus.

Functions of the uterus include nurturing the fertilized ovum that develops into the fetus and holding it till the baby is mature enough for birth. The human uterus is pear-shaped, about 3 inches long, 1.8 inches wide and 1.2 inches thick.

A typical adult uterus weighs about 60 grams. It increases in size and weight while stretching and growing to hold the baby during pregnancy. Just prior to birth, the mother's uterus typically weighs about 1.1 kg, or approximately 18 times the pre-pregnancy weight as it grows to support a baby averaging 3.5kg at birth.

Structurally the human uterus is divided into 4 distinct areas: the fundus, the corpus, the cervical canal and cervix as seen in figure 16.1.

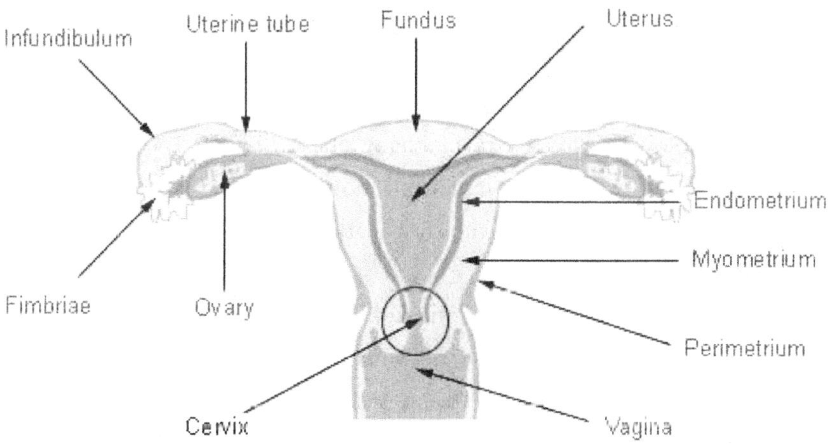

Fig 16.2: Cross-sectional diagram of the uterus, showing the three layers, endometrium, myometrium, perimetrium.

As seen in fig. 16.2 the uterine wall is made up of 3 distinct layers, with the innermost layer (composed of an epithelial and mucosal layer) called the endometrium. The inner endometrium is the functional layer of the uterus where the mucosal part of it gets sloughed off during the menstrual cycle. During pregnancy the endometrium, under the influence of the progesterone hormone, also creates the placenta and umbilical cord. The outer layer is called the perimetrium which is the lubricating layer around the uterus to reduce friction during movement.

The middle layer is the muscular layer called the myometrium. The myometrium is the smooth muscle layer of the uterus. It is the strongest muscle in the female human body by weight capacity and the force generated per cross-sectional unit area. It can stretch and contract with considerable force during childbirth. During childbirth, the uterus exerts 25 to 100 lbs (100 to 400N (Newtons)) of force with each contraction. This is an enormous force acting on the head of a baby as it makes its way past the cervix during birth. To put this force in perspective, it is 3 to 12 times the weight of an 8-pound baby acting on the child's head. The birth process is hence at the intersection of the physics of the forces acting to push the baby through the cervical wall together with biochemistry of effacement (or dissolution) of the cervix itself during birth. The large range of this force also means the uterus under unfavorable conditions can deliver a crushing blow to a baby's head. Variables like the size and weight of the baby, the size of the head, etc., also play a role in how this force acts on a baby's head. A few contractions on the high end of the range against a rigid non-effacing cervix can deliver a devastating blow, resulting in the death of the baby or causing significant brain damage.

The uterus is built strong and muscular exclusively for the birth process, but its strength can be a boon or a bane in the birth process and its characteristics vary widely in the population. The natural contraction response of the uterine myometrial layer is seldom dangerous to the baby, but artificial stimulation can cause an unpredictable contraction response, and is proven to be dangerous to the baby in many circumstances.

The Placenta

The placenta is a special temporary organ that develops during pregnancy inside of the uterus connecting the umbilical cord of the baby to the mother. The function of the placenta is to allow nutrient uptake, thermo-regulation, waste elimination, and gas exchange from the fetus using the mother's blood supply. It also responsible for the production of pregnancy related hormones and provides protection from infection for the fetus.

As seen in figure 16.3, the placenta accomplishes its functions by maximizing the area in contact with the maternal uterus and its rich blood supply using small finger like projections called the chorionic villi. These villi themselves have further substructures called microvilli which help increase the surface area in contact even more. Although they do not come in direct contact, through these structures in the placenta the maternal and fetal circulations meet to exchange gases and nutrients. The placenta also metabolizes a number of substances and can release metabolic products into maternal or fetal circulations. The placenta is also said to play a role in gene expression during pregnancy.

PLACENTA IMPLANTED ON UTERUS WALL

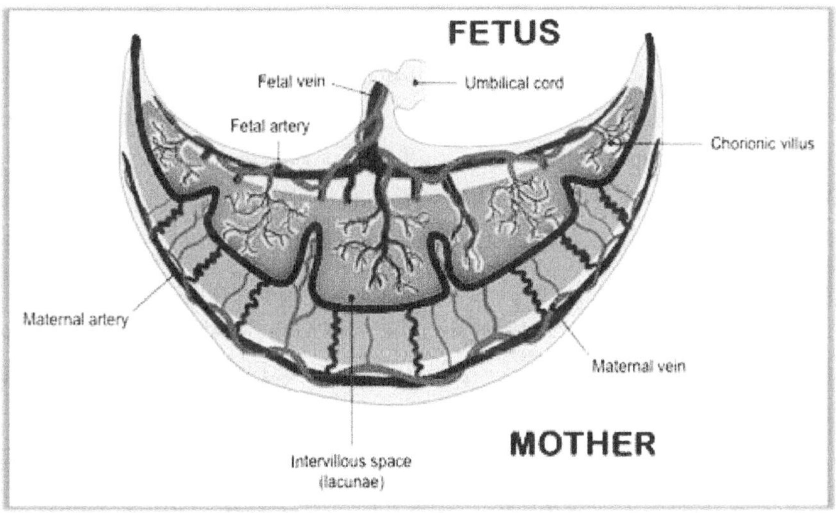

Fig. 16.3: Cross-section of the placenta and uterus wall at the site of implantation, showing the fine capillaries involved in the gas exchange between mother and fetus.

Given all this functionality, the placenta is potentially the most important organ of pregnancy, yet its typically not given due consideration unless there's a problem. Its position, strength of attachment to the uterus, etc., play a significant role in many pregnancy complications. The placenta is designed to work between the systolic and diastolic pressure of the maternal blood to provide its nutrient and gas exchange functionality. This equilibrium is potentially disturbed when labor contractions begin.

When the delicate membranes of the placenta get squeezed by these forces meant to expel the baby from the womb, blood could get squeezed out of it, impacting the gas exchange and the amount of blood reaching the fetus. That's the reason there can't be more than a certain number (2-3) of contractions per 10 minutes. The issue is even more complex after the cushion of amniotic fluid is lost during labor.

The forces acting on the placenta during a contraction in the absence of amniotic fluid are a function of the force of the uterine contractions, the position of the placenta and the position of the baby and the umbilical cord. All these variables impact blood flow and oxygen supply to the baby. While it's typically not a problem with natural contractions, induced contractions (being much stronger) can result in loss of effectiveness of the placenta in keeping the baby adequately supplied. Hence for best outcomes it becomes paramount that once the amniotic fluid is lost, the baby is delivered in a very short amount of time. Because it depends on a slew of variables that can be different in every individual case, medical professionals disagree on exactly how long the fetus can survive without suffering brain damage. They typically point to the possibility of infection as the reason one can't wait for too long, but the reality of the situation is this disruption of blood and oxygen to the baby from the placenta that's the bigger cause for concern.

A Brief History of Induced Labor and Oxytocin

Different societies have evolved their own techniques to get the woman to deliver when a woman crossed 40-42 weeks of gestation with no indication of labor. Techniques varied from herbs and traditional medicines and gentle massaging to the beating of drums and firing of guns (remember the humorous phrase 'son of a gun'?). Certain ancient religions even have their own gods and deities that are dedicated to the pursuit of good birth whose blessings were believed to lead to a positive birth outcome for a pregnant woman.

Crossing this 42-week time limit is known to be a risk to the lives of both the pregnant woman and the baby. Historically, pregnancy and childbirth were one of the primary causes of death for women. Then came the C-section, originally used as a last-ditch effort to cut open the mother to save the baby, and which over time was perfected to be safe for both the mother and child, becoming a life saver for women having difficulty delivering a baby.

In 1906 oxytocin and its uterine contraction property was discovered by British pharmacologist Henry Dale. It is named after the Greek word 'oxytocic' which means 'quick birth'. Oxytocin is a naturally occurring hormone produced by the hypothalamus area of the brain and secreted by the pituitary gland in mammals. It became the first polypeptide hormone to be sequenced and artificially synthesized by Vincent Du Vigneaud who was awarded the Nobel Prize in 1955 for his work.

The modern birth technique of induced labor became possible during the 1980s. The discovery of oxytocin at the dawn of the 20th century combined with advancement in electronics, sensors and techniques to monitor a woman's contractions during labor, fetal heart rate, etc., have given medical sciences the capability to safely deliver a baby without resorting to a C-section when labor doesn't occur naturally. Since the 1980's induced labor has become the de-facto method of managing labor in maternity wards across the US and has been gaining traction in the rest of the world as well.

Interestingly, oxytocin is a molecule that acts both as a neuro transmitter in the brain and as a hormone in other parts of the body. It does not cross the blood brain barrier making this dual functionality possible. The oxytocin from the pituitary gland is the hormone part of it, released into the bloodstream and circulated throughout the body. Inside the brain on the other side of the blood brain barrier, it seems to be produced in oxytocin-projecting neurons separate from the pituitary gland. While it is most commonly described for its hormonal effect of uterine activity stimulation in women, it is common to both genders.

Various scientific studies have concluded that oxytocin is involved in human emotions. It is thought to be released during touching, hugging and orgasm. It also seems to play an important role in social recognition and bonding and is thought to be involved in the formation of trust between people, kindness and general emotional stability. Research has further expanded the involvement of oxytocin to a wide variety of physiological and pathological functions such as sexual activity, penile erection, ejaculation, pregnancy, uterine contraction, milk ejection, maternal behavior, social bonding, stress and many more.

Suddenly oxytocin is becoming this touted cure-all love hormone, and even a cure for some symptoms of autism and schizophrenia due to enabling social bonding. But as any sensible physician would tell you, nobody should meddle with hormones due to enormous consequences they have on the body. The long term effects of artificially meddling with hormones in the human body remains riddled with unknown dangers.

A Brief History of Pitocin

Synthetically produced oxytocin is marketed under the name 'Pitocin' in the US and under various other brand names in other countries. A simple trademark search reveals the Pitocin name itself was registered by Parke Davis & Co, in 1928 long before synthetic oxytocin was produced. Originally, Pitocin was the name given to naturally sourced oxytocin from pituitary glands of cattle gathered from cattle slaughterhouses. The new name was a move to get away from the negative attitudes that had developed toward its predecessor (named 'Pituitrin') which had quickly become the norm to help with labor after 1909 – within 3 short years of oxytocin's discovery in 1906. However, Pituitrin was impure, often contaminated and resulted in adverse outcomes for mothers and babies. Back then women were given one injection of it, and up to 50% of pregnancies where it was administered the result was massive unregulated contractions, uterine rupture and even the death of the child, mother or both. The newly named Pitocin was a more purified form of the same naturally sourced version that tried to address the contamination issues. Yet the mode of administration remained in the form of injections. As there were no good means to measure contractions in those early days, it was not known how long the hormone lasted in the bloodstream after injection and how long they brought on sustained contractions.

With the ability to synthesize Pitocin in a lab beginning in the mid to late 1950s, it was possible to obtain a pure form of the oxytocin hormone. However, its use in inducing labor remained dangerous and its record very spotty at best. While it was possible to induce contractions with it there was not a good way to control and monitor contractions. Due to the dangers involved, for a couple of decades Pitocin was only used as a drug of last resort when all else failed, with C-sections and forceps remaining the birth options of choice depending on where along in the natural birth process progression was halted.

The 1960s and '70s marked an era of enormous progress in modern medicine when many new electronic sensors and other instrumentation came into being. The cardiotocograph or the fetal heart rate monitor was invented by Dr. Orvan Hess in the 1950's and refined further throughout the 60's and beyond. But the birth process was still more centered on "intervention when needed" rather than the managed birth process that it has since evolved into.

Things changed in 1980 when synthetic oxytocin drugs Syntocinon (now discontinued) and Pitocin were approved for induction of labor. Pitocin was approved by the FDA for "Initiation or improvement of uterine contractions, and control of post-partum hemorrhage." It was the harbinger of the managed birth process and the philosophy of "active management of labor" in delivery rooms across the US and the developed world (and catching on very quickly everywhere else). The Pitocin trademark as of 2020 was owned by PAR pharmaceuticals, but having become a generic drug, it is produced by many companies around the world.

Pitocin is considered a miracle drug for its use in stopping post-partum bleeding. The delivery of the baby and the placenta leaves a gaping wound in the uterus at the site of attachment of the placenta to the uterus. All that essential blood supply to the baby developed over the months of pregnancy needs to be quickly reversed, the bleeding stopped and the uterus returned to its pre-pregnancy size and state. Pitocin is capable of doing just that in its post-partum use. It is credited with saving millions of women from bleeding to death after childbirth in delivery rooms across the world, and is listed on the World Health Organization's (WHO) essential medicines list (EML) for this purpose.

But the effect of the synthetic version of the hormone on the mother's body before delivery and its effect on the baby, is a subject of controversy in medical and academic circles. Numerous doctors and nurses have suggested the uterine contractions brought on by the synthetic hormone are much more forceful than natural contractions; so much so that the label on the drug warns against such a possibility in certain individuals. But that has not impacted the popularity and widespread use of this drug, causing unfavorable birth outcomes that remain hidden to this day.

The Induced Labor Process

Synthetic oxytocin or Pitocin is the main anchor of the induced labor process. This refers to the intervention that is done when the natural labor process is stuck or when a pregnant woman is postdated beyond the delivery date without indication of labor. In delivery rooms across the world Pitocin is used for its uterine contraction property in an attempt by modern medicine to replicate the events that happen naturally in cases where birth doesn't seem to be progressing naturally. It is therefore important to look at some of the steps involved in labor in more detail.

Cervix ripening: The cervix plays the very critical role of holding the baby inside the mother as it grows and develops in her uterus. A strong cervix wall at the mouth of the uterus ensures the baby stays in for the full term of the pregnancy and avoids the end result of a premature delivery. As the end of the gestational period arrives, the cervix begins to ripen and dilate to make way for the baby to come out. This dilation process goes hand in hand with the beginning of uterine contractions and labor. The exact mechanism of how this happens naturally is not fully understood. One aspect of natural contractions is that they do not come on when the cervix is not ready, and this is again a natural protection mechanism for the baby inside. A cervix that is soft and pliable is considered ripe for delivery versus when it is hard and rigid, where it is considered non-compliant and unready for labor and delivery.

A rigid cervix is responsible for increased intrauterine pressure during contractions which is associated with reduced uteroplacental blood flow. When inducing labor contractions externally with Pitocin, if the cervix doesn't dilate at the appropriate time in spite of the onset of strong contractions it can result in crushing blows to the baby's head and reduced uteroplacental blood flow with every contraction, resulting in brain injury which later results in developmental disorders or even the death of the baby.

Given such importance of cervical ripening and cervix dilation and a history of neonatal deaths and other adverse outcomes (brain damage) due to lack of cervix ripening, modern medicine offers multiple techniques for achieving cervix ripening. As far back as 1964 Edward bishop set forth a pelvic scoring system to assess the readiness of the cervix for labor.[15.4] It takes into consideration cervical dilation, position, effacement, consistency of the cervix, and fetal station. The score ranges between 0 and 13: a score of 8 or better is favorable for induction, while a score of 6 or less indicates a need for preparation of cervix using the following techniques.

- *Stripping the membrane:* Effective only in a very small percentage of women, this technique involves running an exam finger between the lower edge of the amniotic sac and the internal cervix wall gently separating the two, which might stimulate labor and cervix ripening.
- *Cervical ripening balloon:* This is a more mechanical, non-chemical way of ripening the cervix, where a double-balloon is inflated from both the internal and external sides of the cervix wall. The resulting additional pressure specifically directed on the cervix wall causes the tissue to soften and ripen over time causing the thinning and effacement of the cervix. The balloon is kept inside for up to 12 hours or until it falls out by itself.
- *Ripening agents:* Certain chemical medications are available that can cause cervical dilation in a significant number of women. This includes Cytotec (misoprostol) and Cervidil (prostaglandin E2) which can be administered in the hospital or outpatient settings. They come in tablet form that can be inserted into the vagina, taken

orally or administered in liquid form alongside Pitocin during induction. When done ahead of induction which is the more appropriate way to do it, in postdated women, often, multiple doses over many hours are required to get the cervix ready for labor. Depending on the method used the baby might need monitoring in the hours following the administration of ripening agents.

Amniotomy: This refers to the act of artificially breaking the water or the amniotic sac in a pregnant woman mainly as a way to help labor along. The amniotic sac is composed of 2 layers. The external layer, which is attached to the uterus at the site of placental attachment, is the chorion, and the internal layer in contact with the amniotic fluid is the amnion. The amniotic fluid is comprised of a serous fluid initially, and later added onto through fetal urine as the pregnancy progresses.

Water break tends to happen naturally preceding natural onset of labor or during labor. Amniotomy is known to have has practiced by obstetricians historically for a few hundred years for speeding up labor. In modern times, the use of fetal monitoring equipment requires an amniotomy. The fetal scalp electrode (for accurate fetal heart rate monitoring) and the intrauterine pressure catheter (to measure the strength of contractions used to monitor the baby in a cardiotocograph) do not work without the access provided by amniotomy.

Once amniotomy is complete it is important to deliver the baby within a set time (preferably 8 hours or less). The risk of cord compression, infection, etc., start climbing once the protection of the amniotic fluid is lost. Hence it is a natural bodily response to speed up contractions upon loss of amniotic fluid even without augmentation of labor. The speeding up is due to the positive feedback loop described in Figure 1.2 in chapter 1 due to the direct action of pressure transferred from the baby's head to the cervix rather than the even distribution of pressure that might exist in the presence of amniotic fluid.

Doctors don't seem to agree on a fixed time frame or an exact deadline from the time of water breakage/amniotomy to the time of delivery vaginally or through C-section for best outcomes. Estimates of this time vary widely among the medical community, between 8 and 72 hours. I believe this safe time limit is closer to 8 hours based on the experience of some parents I have talked to. Failure to deliver in that time could result in negative outcomes due to one or more of the following reasons:

- Uneven pressure on the placenta from contractions in the absence of amniotic fluid might result in unpredictable effects on the gas exchange of oxygen supply and carbon dioxide elimination from the baby;
- Uneven and unpredictable pressures acting on the umbilical cord could result in cord kinking and compression which will again interfere with the gas exchange between mother and baby; and/or
- The direct pressure acting through the head of the baby against the cervix during contractions will result in traumatic brain injury that is not easily recovered from if allowed to continue beyond a certain time frame.

Infections can be controlled with antibiotics, but the above effects are unpredictable and easily constitute major risks as the time between loss of amniotic fluid and birth increases. There are cases where babies have been induced with Pitocin that had positive outcomes even though they took many hours or even days, but upon closer scrutiny these are usually instances where the induction was done with the amniotic sac still intact. The entire dynamic changes once the amniotic fluid is lost for the reasons stated above.

Contractions: Uterine contractions are the main mechanism by which the baby is delivered at the end of pregnancy. While these may start spontaneously at or around 40 weeks in most women, both premature contractions resulting in premature birth of the baby and postdating of pregnancy where no contractions arrive even after 40 weeks of pregnancy are fairly common.

Contractions work when the endometrial muscles of the rear fundus side of the uterus contract, thereby propelling the baby headfirst through the cervical canal. Too little contraction force and there is no progress in labor, but too forceful a contraction and there could be head trauma to the baby, uterine rupture, reduced uteroplacental blood flow and even death. The force of the contraction is also a function of how soft (compliant) or rigid (non-compliant) the cervix is. A compliant cervix results in the generation of lesser contraction force, but there is still progress in labor.

The frequency of contractions is a factor that is controlled carefully while inducing labor. A maximum of 5 contractions per 10 minutes is medically allowed (normally 2-3 every 10 minutes). Anything higher and there is overstimulation called tachysystole contractions of the uterus. Another variable is the duration of a single contraction. When a contraction lasts longer than 2 minutes the condition is called Uterine Hypertonus. When tachysystole or hypertonus conditions of the uterus exist (called hyperstimulation) there is not enough time for the placenta to recharge with fresh blood supply for the baby between contractions, resulting in low oxygen outcomes like hypoxia and fetal acidosis. These conditions typically lead to a non-reassuring patterns on the cardiotocograph, which need to interpreted and the baby C-sectioned out.

Thus, any excursions in any of these controlled conditions in induced labor can result in brain damage to the child. So, it is important to monitor these parameters very closely when undertaking medically induced labor. Such monitoring is made possible in current times using a machine called the cardiotocograph, more commonly known as an electronic fetal monitor (EFM). It monitors both the fetal heart rate and the occurrence of uterine contractions in the mother.

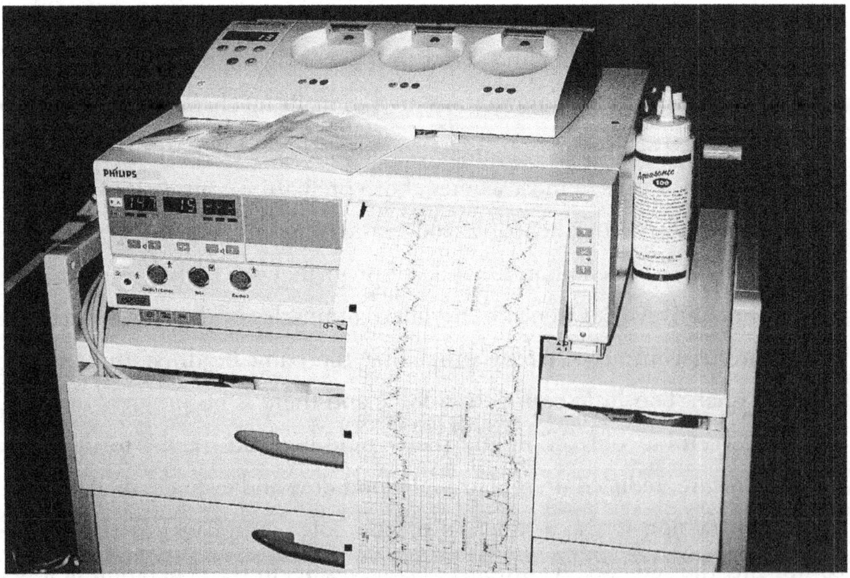

Fig 16.4: Typical cardiotocograph seen in delivery wards across the world.

The availability of electronic fetal monitoring is the main reason induced labor is even possible. While this is a valuable tool in the hands of the OB/GYNs and maternity ward nurses, its interpretation is a subject of controversy as well. Court cases have been won and lost over their outputs and the correlation with the outcomes for babies.

Electronic fetal monitoring: Electronic fetal monitoring is based on the principle that the fetal heart rate goes down during a contraction and recovers quickly after the contraction abates. If there is lack of swift fetal heart rate recovery after every contraction then the fetus is said to be in distress and doctors might force a C-section if the readings from the fetal heart rate (FHR) monitor show signs of fetal distress. However, the actual interpretation is a subject of controversy because of the possibility of errors in the instrumentation placement, positioning, quality of the signal, fetal position and so on. Initial monitoring before any cervical dilation is done with an abdominal monitor, and when the cervical dilation reaches 2cm or more then the head of the fetus is reachable and a fetal scalp electrode is placed on the scalp. Yet while the accuracy of readings is improved, the interpretation of what they mean and what really constitutes fetal distress is a totally different ball game.

Interpretation of monitor readings is a complex undertaking, and scientific papers continue to be published trying to bring clarity and accuracy to the process. The key to successful monitoring is to establish a baseline fetal heart rate, which is easy to do when contractions are few and far apart. The normal fetal heart rate is between 110–140 beats per minute. An abnormal condition is said to exist when the baseline heart rate is not within this range. Bradycardia is when the baseline heart rate is less than 110 bpm, and tachycardia occurs when the baseline heart rate is greater than 140 bpm. Each of these states is indicative of certain conditions like fetal heart issues, fetal malposition, maternal fever, fetal hypoxia, fetal anemia, amnionitis and many others which are cause for concern.

Doctors have created a three-tiered fetal heart rate interpretation system. Categories I and II are considered safe while category III represents fetal distress. The latter is mostly characterized by decelerations in heart rate from baseline between contractions.[2] Understanding FHR flows may be beyond the understanding of typical parents, but delivery room doctors and nurses are trained for years to pay close attention and interpret those readings with their training and experience. They are expected to accurately identify fetal distress and other conditions, which they mostly do. But in instances where they don't act appropriately at the right time can end up causing irreparable brain damage, which can range from mild ADD issues all the way to quadriplegic cerebral palsy and even fetal death. Time is of the essence in some of these issues. Imagine what would happen to a person that suffocates for 10 minutes unable to breathe. They go into a coma from brain damage or die. That is exactly what could happen to a fetus in distress.

There are significant flaws in the current state of induced labor monitoring. Internal fetal heart rate monitoring is typically used during the inducing of labor. Fetal heart rate monitoring tends to give attending OB/GYNs the confidence that everything is going well during induction of labor even if that might not be the case. The basis of fetal monitoring is the belief that hypoxia or ischemia will cause fatal heart arrhythmias that can be detected and acted upon. When the brain is not receiving sufficient oxygen in its autonomous control regions of the brain stem the initial response is that it sends signals to the heart to beat faster to increase blood flow. But when the lack of oxygen persists it impacts the brain's ability to send out those signals and the heart rate might fall from baseline. At that point a lot of damage might have already occurred, particularly to neurons in the lobes of the cerebrum. A vast majority of babies C-sectioned out as a result of fetal distress are known to suffer from neurodevelopmental disorders, all of which fall under the umbrella of autism spectrum disorders.

The other issue is the traumatic injury that can be inflicted by Pitocin induced contractions. It is a known fact that the brain doesn't sense pain due to damage to itself, unlike the way it can sense damage to other parts of the body. Electronic fetal monitoring has absolutely no way of monitoring traumatic injury outside of measuring the pressure of contractions. Fetal head compression and intracranial pressure are impossible to accurately measure. Upon loss of amniotic fluid this pressure is all directed at the fetal head trying to make its way past the cervix in the second stage of labor. Unfortunately, traumatic brain damage is not a parameter electronic fetal monitoring can actually warn against. Obstetricians that ignore this time aspect end up causing brain damage in babies unlucky enough to come under their care. Unfortunately, there are no signs of fetal distress and so the fetal monitor in these cases actually provides a false notion of safety.

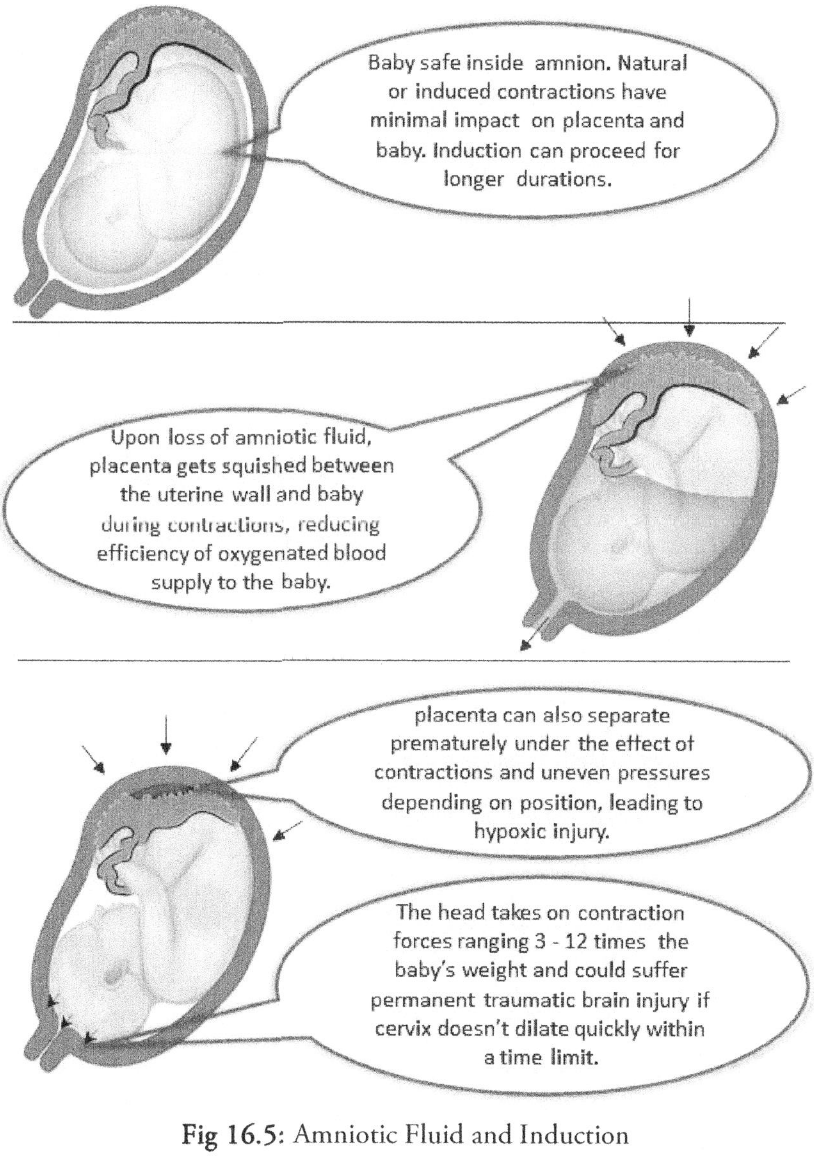

Fig 16.5: Amniotic Fluid and Induction

Table 16.1 Contrasting two kinds of induction.

Two Types of Pitocin Induction	
With Amniotic sac intact	Post Amnion rupture
Uniform distribution of pressure inside the amniotic sac.	Lack of fluid medium means pressure is non-uniform across various parts of baby and placenta.
Pressure from contractions transmitted to cervix through the amniotic fluid.	Pressure from contractions transmitted to the cervix through the head of the baby.
Pressure on the umbilical cord and placenta is fairly uniform and there is little risk of hypoxia or other damage.	Risk of hypoxia and ischemia from uneven placenta compression and cord compression is heightened.
When people say I was induced for 30 hours and still have a fine baby, this is the type of induction they are referring to.	This sort of induction for more than 4-6 hours (depending on contraction frequency and other variables) causes sure shot traumatic brain injury.

Pitocin administration: Pitocin is the drug of choice for inducing and controlling contractions. When a pregnant woman is postdated for delivery or when natural labor is stuck, delivery rooms default to using Pitocin for active intervention and management of labor. Most hospitals and obstetricians develop and follow their own methods of how induction is done. In a postdated woman the most commonly followed sequence of interventions would be as follows:

1. Cervix ripening
2. Pitocin administered
3. Amniotomy (if needed)

However, depending on the individual practitioner, the choices of cervix ripening time and technique, amniotomy timing, etc., may vary. In spite of all the monitoring and available techniques, obstetrics is still more of an art than science, and the experience, expertise, judgment and beliefs of the specific obstetrician determine the techniques followed and the outcomes achieved. As they always say, no two deliveries are exactly same.

Pitocin only lasts in the blood stream for three to six minutes, so when used for labor induction it is administered as an intravenous (IV) fluid. Nurses typically adjust the dosage within a set range to try and achieve the desired rate of contractions. Once the desired rate of contractions is achieved, they continue to monitor progress of cervical dilation with time. Induction works best when the cervix is ripe or already dilated some when induction is started. The fetal heart rate and any signs of fetal distress is then carefully monitored and any signs of fetal distress are addressed immediately with an emergency C-section. If progress of cervical dilation is not rapid enough after the start of Pitocin induction, induction needs to be stopped and a C-section needs to be performed.

Failure to perform a C-section at this juncture is unfortunately the biggest cause of medical malpractice in the US and the cause of autism in many children (including my own). This is also the Pitocin contra-indication described in the FDA approved Pitocin labeling comes in: "Where adequate uterine activity fails to achieve satisfactory progress." Numerous obstetricians who believe blindly in the virtues of Pitocin and have not taken the time to study the warning labels and tend to increase the Pitocin dosage rather than stop induction upon such contra-indication. In cases of low risk healthy mothers and babies, there is typically no fetal distress, which is the only other reason for them to stop induction. OB/GYNs typically tend to proceed all the way in their zeal to avoid the C-section, causing irreversible brain damage as indicated in the Pitocin labeling.

Is Pitocin the only game in town to induce labor? Apparently not. There are other natural ways of inducing labor as described by the Global Library of Women's Medicine (GLOWM), the educational arm of the International Federation of Obstetrics and Gynecology (FIGO). On their site they describe the following non-medical means of inducing labor and cervix dilation:

- Sexual intercourse
- Breast stimulation
- Herbal preparations

- Homeopathic solutions
- Purgatives
- Enemas
- Acupuncture
- Stripping of the membranes

Expert OB/GYN's typically use some of these techniques like enema and striping of the membranes the night before the date they attempt induction, so as to make the cervix compliant, and it makes a world of difference to the final outcome for the baby.

Key Take Away: The movement towards active management of labor in hospitals has led to medical malpractice involving abuse of the induced labor drug Pitocin. There exists a major flaw in fetal monitoring during induced labor. Electronic fetal monitors only show hypoxic injury, while traumatic injury can happen, when induction is carried out against a non-compliant cervix and such injury will go undetected by the monitors. Use of Pitocin past the point of contraindication indicated on the label has led to traumatic brain damage in babies, which is later called autism. Damage at contra-indication is particularly bad as induction is done after an amniotomy.

CHAPTER 17

Looking Past the Smokescreen

Like in most medical practice, OB/GYN's study their subjects in various colleges. There is not a fixed central authority which directs the knowledge and practice of Obstetrics and Gynecology in the US. Like every other medical field, doctors are trusted to read through the warnings of the medications they use and exercise their best judgement on their use as they see fit. There's the American board of Obstetrics and Gynecology (ABOG), an organization that tests knowledge of the subject and provides board certification, but OB/GYNs do not need to be board certified to practice. Then there is the American Journal of Obstetrics and Gynecology (AJOG) that puts out the latest studies and scientific papers and advancement in the field of Obstetrics in the US Many of these papers (as is common in this field) point to 'unknown' origins of autism. There are other international organizations like the International Federation of Obstetrics and Gynecology (FIGO) who provide resources to educate doctors and nurses, but sometimes common sense suggestions coming from international organizations tend to get written off in the US as more suited to 'low income' countries.

The American College of Obstetricians and Gynecologists

The closest an organization comes to providing centralized practice guidance information in this field is the American College of Obstetricians and Gynecologists (ACOG). They have useful information on obstetrics including labor induction on their website acog.org. But even the ACOG guidance directives on inducing labor has been fairly unclear from what I understand looking through their site and publications. They speak of a C-section as the undesirable outcome of failed induction to be attempted only in the case of fetal distress. There seems to be no mention of Pitocin contra-indication anywhere in ACOG literature as of early 2021. Their lack of clear guidelines regarding Pitocin contra-indication has potentially resulted in creating a class of OB/GYNs who engage in demonizing of C-sections, exaggerating the negatives and disregarding the necessity of timely C-sections in cases of Pitocin contra-indication. These OB/GYNs "excel" in hammering babies out with increased doses of Pitocin and are to blame for the situation with respect to autism from medical malpractice (what I call Aumalism) in the US.

Fig 17.1 is a screenshot from the ACOG website showing their guidance on labor induction. They include fetal death as an outcome from induction and they blame it on maternal conditions. Fetal death particularly during birth happens mainly as a result of hypoxia, ischemia or head trauma. All these conditions (if not resulting in immediate death of the baby) can lead to neurodevelopmental disorders. This is particularly true in the case in healthy mothers and babies. That is the intermediate outcome that they seem to not acknowledge in any of their more detailed papers or guidelines. They talk about induction not being effective, but do not seem to elaborate anywhere the circumstances during induction when they should deem induction ineffective and resort to a C-section.

The ACOG should have been the first people to link neurodevelopmental disorders to birth injury and the word autism, but they seem to have avoided the topic altogether, leading one to think it's an obvious conflict of interest and they are perhaps trying to protect their practitioners from potential lawsuits for mistakes from their past. Most of their experts seem to be enraptured postulating about the chemical effects of externally administered oxytocin on the fetus, faulty oxytocin signaling pathways, alteration of fetal oxytocin receptors, etc., as the cause of autism when the actual cause is physical trauma to the brain from forceful Pitocin contractions against an unripe cervix – when obstetricians pursue induction contrary to the instructions on the label of FDA approved Pitocin.

https://www.acog.org/womens-health/faqs/labor-induction

What are the risks associated with labor induction?

With some methods, the uterus can be overstimulated, causing it to contract too frequently. Too many contractions may lead to changes in the fetal heart rate, umbilical cord problems, and other problems. Other risks of cervical ripening and labor induction include the following:

- Infection in the mother or fetus
- Uterine rupture
- Increased risk of cesarean birth
- Fetal death

Medical problems that were present before pregnancy or occurred during pregnancy may contribute to these complications.

Is labor induction always effective?

Sometimes labor induction does not work. A failed attempt at induction may mean that you will need to try another induction or have a cesarean delivery. The chance of having a cesarean delivery is greatly increased for first-time mothers who have labor induction, especially if the cervix is not ready for labor.

Fig 17.1: Screenshot from the ACOG website.

In fact, the ACOG has published a paper on their website on "Labor induction or augmentation and autism," a committee opinion originally from the year 2014 and reaffirmed again every couple of years up to the present. In it they link the entire topic of the Pitocin-autism relationship to a chemical effect. The committee seems to be conveniently forgetting that Pitocin has a history of brain trauma and death in its past and its use was banned before being re-approved after the advent of electronic fetal monitoring.

Labor induction or augmentation and autism. Committee Opinion No. 597. American College of Obstetricians and Gynecologists. Obstet Gynecol 2014;123:1140–2.

The link to it on the ACOG website is as follows. https://www.acog.org/Clinical-Guidance-and-Publications/Committee-Opinions/Committee-on-Obstetric-Practice/Labor-Induction-or-Augmentation-and-Autism

If that wasn't bad enough, they don't stop there but instead go one step further by suggesting that all autism is genetic. It's not like there's been a dearth of scientific papers citing the risks in active management of labor. There's also been papers highlighting the risk of malpractice when it comes to labor management. What's been missing is the tagging of what these papers call 'adverse outcomes' with the term 'Autism.' Below is a small sampling of science papers and studies that have talked about malpractice risks from as far back as 1987, within a few years of the official approval of Pitocin for inducing labor in 1980 by the FDA for cases of necessity:

1. Nocon JJ, Coolman DA. Perinatal malpractice. Risks and prevention. J Reprod Med 1987;32:83–90.
2. Oláh KSJ, Steer PJ. The use and abuse of oxytocin. The Obstetrician & Gynaecologist 2015;17:265–71
3. Gregory SG, Anthopolos R, Osgood CE, Grotegut CA, Miranda ML. Association of autism with induced or augmented childbirth in North Carolina Birth Record (1990–1998) and Education Research (1997-2007) databases. JAMA Pediatr 2013;167:959–66.
4. Smallwood M, Sareen A, Baker E, Hannusch R, Kwessi E, Williams T. Increased Risk of Autism Development in Children Whose Mothers Experienced Birth Complications or Received Labor and Delivery Drugs. *ASN Neuro*. 2016 Aug 9;8(4):1759091416659742. Doi: 10.1177/1759091416659742. PMID: 27511908; PMCID: PMC4984315.

Each of these articles cite about 30 other articles each, so there's no scarcity of opinions and studies supporting the argument against Pitocin and highlighting the possibility of malpractice. Unfortunately, people in positions of influence in the OB/GYN community have chosen to ignore these voices perpetuating malpractice and autism (Aumalism).

Sometimes it is possible to miss the point even after reading some of these papers, even if you agree they all are solid evidence that perinatal malpractice might play a major role in causing autism. That's because common folks may not make the connection in words used in scientific literature which actually translate to autism in common parlance:

Adverse outcomes = Neurodevelopmental Disorders = Autism

It makes a world of difference reading through the above papers to your understanding of the causes of autism once you make the above connection.

Of course, to their credit, the ACOG have apparently issued guidance to hospitals to come up with their own protocols to follow in cases of Pitocin contra-indication, and many hospitals do have such protocols in place, but in many cases, the OB/Gyns handling the cases have free reign over what they do and ignorant OB/Gyns continue to destroy babies, by crossing that limit, almost every time it happens. In many ways, the void left behind by organizations like the ABOG and ACOG that have been vague about guidelines regarding Pitocin contra-indication has resulted in medical insurance companies taking the lead to demonize the C-section and perpetuate information extolling the safety of Pitocin. Such behavior by insurance is also a potentially secondary factor to blame for Pitocin abuse.

The Biased and Active Management of the Messaging

Everybody who's followed the news cycles in recent times knows how it's possible to manage the messaging online by hiring public relations or online reputation management firms. In fact, many of these firms have gotten online image management down to a science. The messaging of the safety of Pitocin and its role in causing autism, unfortunately, seems to be one of the pioneering cases of such active message management. The first thing that I noticed about online forums, is there seems to be a horde of folks willing to go ballistic on any parent casting a genuine doubt on the safety of Pitocin induction based on what they are seeing with their kids. There is this diversionary tactic that seems to be a very deliberate (and thus far hugely successful) attempt to confuse the issue with other possible causes, unknown diseases and everything else they can think of, all while vouching for the safety of Pitocin, citing supposedly hundreds of cases of successful inductions. Here I was, looking at these forums, reading these deceptive messages, while sitting with solid evidence of an MRI scan and now clear recollection of the events that lead to my children's condition.

It is clear that there is very active management of the messaging about Pitocin by a public relations firm or a legal organization. There are deliberate efforts to create confusion in the minds of people in multiple online forums. The search engines are being manipulated by some internet reputation management firms such that the pages supporting Pitocin's safety rank higher than any others that raise questions about its safety. Reading through some of these forums I have seen the same opinion echoed by many impacted parents in such forums, who find themselves shouted down by choruses of what looks like hired hands.

Search for Pitocin Autism on Google and you will see what I mean. Figure 17.2 is a screenshot of a Google search for the words Pitocin Autism. See what came back in early 2017. Nine out of the top 10 search results vouch for the safety of Pitocin. Trying the same search 3 years later in 2020 brings back a few different results. However, it's the same number of results in support of Pitocin, i.e. 9 out of 10.

The Bing search engine from Microsoft brings back as its top result in early 2020 a blog by name autismsd.com from an anonymous author with the picture of a vial of Pitocin for added effect, forcefully vouching for its safety by someone claiming to be a biologist (Figure 17.3). If you have read through the previous chapters you know that the safety of Pitocin is not a question a biologist could answer, but one for a biomechanics expert to speak to. Only the latter can speak to the kind of contraction forces crushing a baby's brain and the placenta in the absence of the protection of amniotic fluid causing traumatic/hypoxic damage on a baby's brain. This is another example of how an innocent blog from a well-meaning blogger is being manipulated by special interests. There is no way a blog like that could rank on top except for some expert manipulation of search engine optimization by some hired hand.

Fig. 17.2: Google search result for 'pitocin autism' in 2017

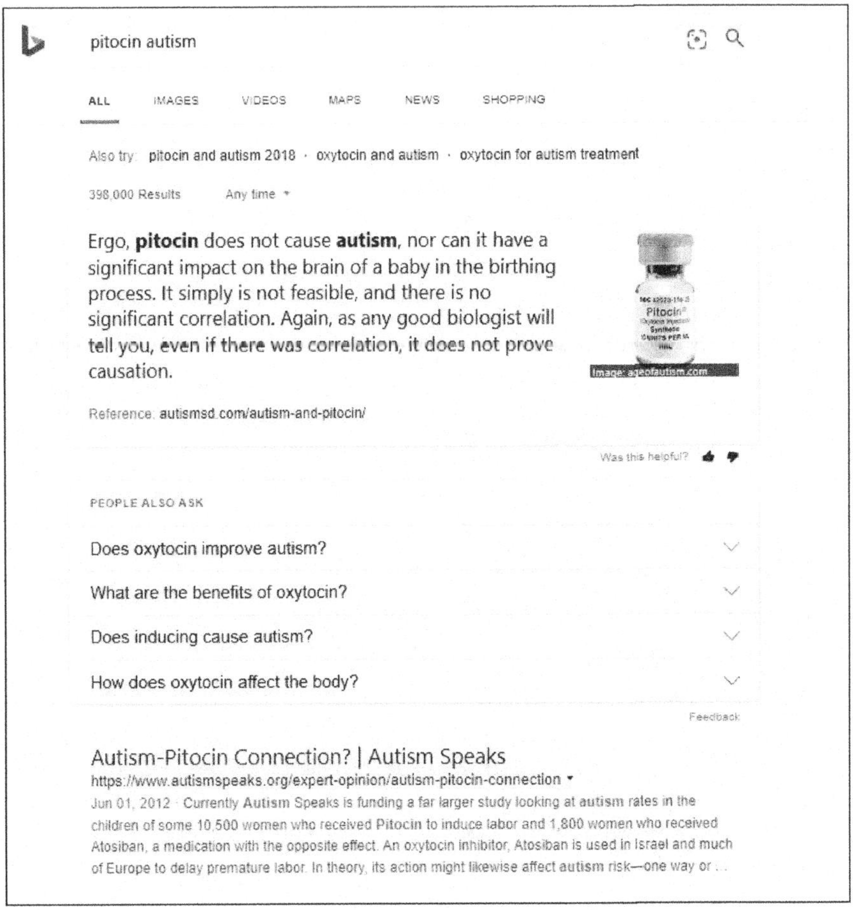

Fig. 17.3: Pitocin Autism search results on Bing in 2020.

As you see in the screenshot, search engine optimization techniques have been applied to internet search engines to have the flawed research vindicating Pitocin ranking higher, even using links from none other than the foremost organization for autism (autismspeaks.org). Their credibility is being abused to further the Pitocin safety argument, contrary to and ignoring everything mentioned on the FDA sanctioned Pitocin labeling. I haven't been able to find a single blog or forum out there where a parents' genuine comment casting doubts on Pitocin's role in Autism is not being diverted or outright discredited with a fake fact from an egregiously fake account. Here's an example in Fig-17.4 from the same anonymous blog ranking first on the Bing search. One parent complains about Pitocin causing her child's autism and another comes by 2 years later to claim genetic predisposition to autism when no such thing has ever existed or been proven. The second account obviously throwing out some rubbish probably done after this blog started ranking higher in the searches. This is just one example; this is the case universally on the internet. It is hard to find any high-ranking blog where legitimate Pitocin concerns are not being misled or drowned out in a flurry of lies.

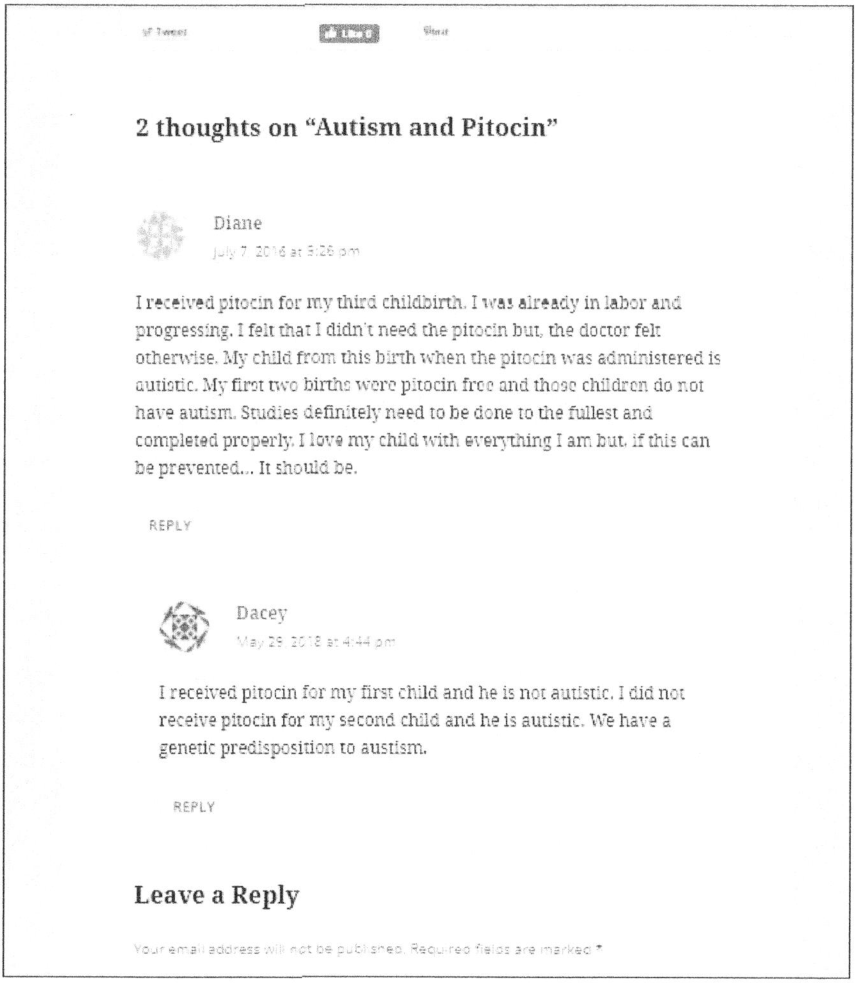

2 thoughts on "Autism and Pitocin"

Diane

July 7, 2016 at 5:26 pm

I received pitocin for my third childbirth. I was already in labor and progressing. I felt that I didn't need the pitocin but, the doctor felt otherwise. My child from this birth when the pitocin was administered is autistic. My first two births were pitocin free and those children do not have autism. Studies definitely need to be done to the fullest and completed properly. I love my child with everything I am but, if this can be prevented... It should be.

REPLY

Dacey

May 29, 2018 at 4:44 pm

I received pitocin for my first child and he is not autistic. I did not receive pitocin for my second child and he is autistic. We have a genetic predisposition to austism.

REPLY

Leave a Reply

Your email address will not be published. Required fields are marked *

Fig 17.4 One example of legitimate Pitocin concerns being shot down with rubbish in online blogs.

Thankfully beginning in late 2020 a paper titled "Increased Risk of Autism Development in Children Whose Mothers Experienced Birth Complications or Received Labor and Delivery Drugs" by Smallwood et. al[17.3] from the Neuroscience Program, Trinity University, San Antonio, TX, USA, have started ranking higher on Google searches, a good 4 years after publication in 2016. Their data is from parent surveys and another dataset from the Southwest Autism Research and Resource Center (SAARC - autismcenter.org San Antonio TX / Phoenix AZ). Their study is hard to refute, considering that it is based on numbers and numbers don't lie. But the facts are that Autism is a consequence of Pitocin-abuse not Pitocin-use and chances are there will soon be a counter study from someone claiming safe Pitocin numbers. Unfortunately, that is the state of studies these days and short of concerted action from impacted parents, even such evidence often falls by the wayside when faced with the true complexity of the facts involved in Pitocin abuse. It is the length of time of induction after amniotic rupture that truly counts, and the correlation is clear in the cases of the healthiest mothers and the contra-indication is defined in the Pitocin label, but it is hard for scientific studies to collect that kind of data, without creating a whole movement around it, like some of us parents are trying to do with the AutismPI.org data collection initiative.

Synthetic oxytocin is now a generic drug that is manufactured around the world by many big and small pharmaceutical companies in the US and in other countries. Some of the bigger names include Pfizer and Novartis. Cheaper generic pricing has meant most manufacturers have razor-thin margins and make little or no money on it. Some have decided to stop manufacturing it, resulting in shortages for what is a widely used drug all over the world for its post-partum use in stopping blood loss after delivery. Manufacturing has also moved to cheaper destinations outside the US. The manufacturer's liability for the product is limited to their accountability to physicians and warning them of the side effects.

The other applications of the drug outside of induction, including abortions and stopping post-partum bleeding, ensure a very steady demand for Pitocin. It's hard to imagine manufacturers or even large suppliers being involved in any internet messaging scam to promote its use in induction, although they are the ones with the kinds of budget to do it. So the possibility cannot be totally ruled out.

The other major party that might stand to benefit from not revealing medical malpractice is the malpractice insurance companies that insure OB/GYNs. They are probably aware that it could be a challenge to their survival if they are hit with large numbers of malpractice suits against OB/GYNs they insure. Compensation in these types of cases could run into multiple millions per case as it comes down to what is fair compensation for ruining someone's entire life starting at birth and the damage caused to their entire family. Malpractice insurers are also the people that will be aware of birth injury lawsuits from Pitocin abuse in extreme cases, where the birth injury was caught immediately after birth and ended up paying out compensation. When they settle out of court, they tend to make parents sign nondisclosure clauses, shutting them up from disclosing the possible pathway to autism.

All this active management of the internet messaging makes one wonder who might be behind all this disinformation and what their objectives are. The involvement of OB/GYN practitioners who have made totally flawed recommendations in the past could be suspected. They might themselves be guilty of medical malpractice, by way of abusing Pitocin and could be very much aware of the damage they have caused. It is in the interest of these medical practitioners to not get sued for their past mistakes. It's also in the manufacturers /marketer's interest to hide the unsafe side effects of the drug, particularly when the cause can be easily hidden. Pitocin itself is not a major moneymaker for big pharma and most manufacturers are out of the business, but sometimes it's amazing how much influence even small players can have on the messaging when it comes to the internet. The malpractice insurers have everything to gain from keeping the causes of autism from Pitocin abuse under wraps. Such attempts at suppression of the truth on the internet could be the result of collusion by some or all of these actors. Whatever it is, it seems to be a devious conspiracy that has potentially kept the truth hidden from the public for decades.

The outcome is one where those involved in the deception are riding a tiger, afraid to let go for fear of retribution, while the crime continues to be committed by some OB/GYNs who in their zeal to avoid the 'intrusive' C-section continue to promote a drug to patients as 'safe and natural' when in fact it is causing irreparable damage to millions of babies over the years in the US. Unfortunately, this 'modern' method of child delivery is spreading rapidly in other parts of the world causing autism there as well.

While Pitocin can be a useful tool in the hands of an expert OB/GYN, there are many others in whose hands it has manifested as this dangerous weapon causing irreversible brain damage, which when not caught immediately is later labeled autism. Most experts would accept these dangers of Pitocin in private, and express their concerns from time to time, but their voices are muted in the face of this massive messaging to the contrary about 'autism'.

Key Take Away: There has been a certain level of deception or withholding of the truth in what the general public are fed regarding the active management of labor in hospitals, concealing medical malpractice involving abuse of the induced labor drug Pitocin. There seems to be an active effort on the internet to manage the messaging, from some unknown actors.

CHAPTER 18

Practical Difficulties with the Legal Case

Any legal case against a medical practitioner is typically an uphill battle, but it is made particularly difficult when the consequence of a doctor's indiscriminate use of a drug is actually a spectrum of disorders. Because there are other factors that can also cause similar effects the cause-effect connection for autism has remained obscured for years. Only the most egregious medical blunders ever see the light of day in court and it is even rarer for victims to emerge victorious in a court of law. Years of Tort reform and aggressive efforts to curtail frivolous litigation has also meant it is much more difficult for legitimate but complex cases like malpractice leading to autism to succeed easily. In fact, it's much easier to sue for injury in an auto accident or an injury at a business/place of work than to sue for medical malpractice.

An Uphill Battle

While impacted parents will readily agree with me, the facts that I have presented in this book will need to gain acceptance in society in general and perhaps a victory in court would help. J.B. Haldane British geneticist, biometrician, physiologist and promoter of science said this about the four stages of acceptance of any truth in society:

Stage 1: ***This is worthless nonsense***
Stage 2: ***This is an interesting but perverse point of view***
Stage 3: ***This is true, but quite unimportant***
Stage 4: ***I always said so !***

I predict anyone other than impacted parents reading this book will go through these 4 stages on their way to accept the ideas put forward in this book. Having been indoctrinated through years of misinformation, it is an uphill task for people to believe that the medical system has failed them so miserably. Since there have been a quite a few successful induced labor births (particularly cases where there is no contra-indication), there is bound to be people out there that will vouch for its safety as well. What is needed is the emergence of the overwhelming proof of brain damage caused by induced labor, which exists but is buried by the misinformation that has come to pervade contemporary beliefs. In a world where deliberate confusion is being sown it is important to prove this clearly so medical practitioners that have committed this blunder of Pitocin abuse can be brought to justice rather than continue perpetuating their crime.

Also problematic are the numerous cases in court against the Federal government working to tie vaccines to autism. While I agree that vaccines are part of the puzzle, my contention is that there needs to have been a complex birth story like high blood pressure, premature birth, hypoxia, meconium issues or Pitocin abuse causing brain injury that precedes the damage that vaccines do. As I have explained, live vaccines have the effect of stopping recovery or cause retrogression of development where prior brain damage is present.

Some conclusions can be drawn from the fate of the cases against autism thus far, many of which have been against vaccines. Of the very few cases filed against vaccines, even fewer are ever successful. The numbers are so bad that the number of victories can be considered next to nothing. Approximately one in a million doses of vaccine distributed results in a successful claim. That is simply because most kids are administered the same vaccines and don't get autism. The case against vaccines is very underwhelming when it is up against such stellar success numbers.

The vaccine injury program has also been very successful in fighting off cases when it comes to proving complaints in federal court. Most cases end up being settled out of court based on past judgments in previous cases. Why is this the case? It is hard for parents to come up with proof that would be acceptable in a court of law. In most cases there's very little evidence of the vaccine causing the autism condition, and in cases where there is, it's hard to pin it down as the lone reason. This is because there is truth to the fact that vaccines are not the sole reason for a child's autism. As I have proposed, vaccines are only the secondary reason in a double whammy that is the true cause of autism.

Then there's the fact of the hospital paperwork. Most parents were probably required to sign a waiver in the hospital – what they call a "consent to treatment" which is essentially a disclaimer that says it's all on you if something goes wrong. Nobody should ever be made to sign paperwork they don't understand, particularly in a hospital setting for something so 'normal' and natural as childbirth. Unfortunately, this is the current state of affairs in every hospital environment in America. Patients are routinely made to sign papers absolving doctors of responsibility for treatments, even in cases where the doctor's knowledge, judgment and skill is called for and makes a difference in the eventual outcome for the patient.

OB Gyns have abused Pitocin indiscriminately for ages. Sometimes it's been administered as a matter of convenience to get labor moving at a faster pace. A lot of these kids do not end up with an autism diagnosis. The contra-indication limit is pretty clear, but the uterine activity is subjective. The vast variety of circumstances in many cases is pretty challenging to solve. But the evidence is there once you put all the pieces together.

Added to the challenge is the fact that the statute of limitations is ticking. It makes sense to fight these cases in federal court as most states have a shorter time frame when it comes to how much time can elapse from the point at which the medical malpractice occurred to the when the case can be filed. Some states are sensible and allow 18 years in the case of children, but many have been overrun by tort reformists who in their pursuit of medical cost reduction have foolishly targeted case age limits for children and made it a lower amount of time like 8 or 10 years. That leaves many of us little choice but to take the issue up in federal courts that allow 18 years before the statute of limitations expires.

A further obstacle is the fact that lawyers don't like fighting new uphill battles. Pitocin abuse litigation has all the potential to be a high stakes legal battle with the potential for billions in payouts across the thousands of cases. There will be big lawyers on the side of malpractice insurers and possibly the drug makers, leading to high legal expenses on the plaintiff's side, long winding arguments, and legal technicalities and pitfalls and traps that have nothing to do with the subject of the case at hand. This is a very technical case with a lot of intricate yet varied technical detail. Typical birth injury lawyers wouldn't take up a case like this on the side of an impacted parent unless they know the plaintiff has the financial and mental staying power to live through the tribulations of a long running case. Plus, many birth injury lawyers have tried to implicate induced labor with no success, which has left a bad taste in their mouth. Most lawyers in this kind of medical injury situation work on a contingency fee basis which makes any case that is less than easy – a long shot for them to even attempt it. They prefer cases with precedent that they know they can win, particularly when they fight on a contingency fee basis. Obviously, they are a business that needs to make money and it wouldn't be worth it for them to take up cases without the likelihood of a victory, which has been elusive with autism cases until now. But as the saying goes "Absence of evidence is not evidence of absence." Proof exists and at autismpi.org there is a data collection effort to bring this evidence out.

The medical establishment and manufacturers have put out a smokescreen that needs to be cleared for the evidence to become visible. While the initial case is against the Autism producer OB/GYNs, the evidence is clear that the drug Pitocin is too dangerous for use in a certain segment of the population where the uterus does not efface. The drug manufacturers have obviously failed to adequately train medical practitioners in its use and hidden the warning from them for the most part. There is clearly a serious conflict of interest, as why would doctors use this drug if they were educated on its long list of warnings.

Medical Malpractice Insurance

In the case of Pitocin contra-indication, at this time any compensation would have to be paid by medical malpractice companies of the insured practitioners. If they paid anywhere close to real compensation for a Pitocin contra-indication case, America's malpractice insurance companies are certain to all go bankrupt. The average successful vaccine autism claim is typically settled by the US Department of Human and Health services for a million or more, with annuities running to 10 million or more over the lifetime of the patient. Just for the sake of numbers, say there are half a million claims of Pitocin abuse of the 3 million suffering from autism. Even a tiny and unjust two million-dollar compensation per individual would require a trillion-dollar payout –bigger than the value of most companies out there and a significant number even when compared to the annual US GDP. The sheer enormity of the stakes involved makes winning what's an obvious case so much of an uphill task.

Claims of medical malpractice against a doctor are mostly settled by malpractice insurance companies as per the limits of liability set by the policy carried by the insured doctor or their practice group. Usually it is not possible to sue the hospital itself unless the doctor was an employee of the hospital. In most cases OB/GYNs are independent contractors with their own practice that use the hospital facilities for birth services. Depending on how much they were insured for, compensation payable by these companies might be limited and not commensurate with the damage done in case of autism. Going after the personal wealth of the Doctors may or may not be fruitful as well.

The hospital may be liable for not keeping doctors from using their facilities, following their policies and for informing the patients of the risks, but that again might be more complex to prove negligence with many OB/GYNs involved and could be a secondary approach once the offending OB/GYNs are singled out. However, when cases are settled by insurance companies out of court, they often do it without admission of guilt and nondisclosure conditions, which make it possible for the perpetrators to continue their practice with no consequence for their actions.

The Role of Medical Boards

State medical boards are government agencies empowered to investigate complaints about doctors and when warranted take action against them. The state medical board is tasked by the state legislature with granting licenses, administration of exams and oversight of medical practice in general within that state. Physicians are licensed to practice in a state by that state's medical licensing board. The federation of state medical boards is an organization that helps the 70 different medical boards (including the 14 osteopathic medical boards) in the US by providing guidelines, standards and best practices for the boards themselves.

Medical malpractice judgments against a physician in court may or may not have an impact with a state medical board depending on which state the ruling is in. Interestingly, even in the same state, in spite of the fact that courts on many occasions take the advice of the boards, the board may not agree with the court's decision. While the court can order malpractice compensation, it's the boards that regulate whether the practitioner continues his practice. Thus, even doctors with proven medical malpractice cases with the court system can continue to practice in many jurisdictions unless such judgments reach abnormally high proportions. Disciplinary action against a physician by a state medical board can be much very consequential since boards have the power to revoke their license to practice, impose fines not covered by malpractice insurance, etc. It is also easier to lodge a complaint against a physician with a state medical board, but harder to prove malpractice as the boards are not typically swayed by just a single report of negligence as such.

The boards are typically comprised of distinguished physicians appointed by state governors and confirmed by state legislature. Being physicians themselves, it's understandable that they tend to be more understanding of other physicians and tend to come down harshly only on the most egregious of malpractice. Since Pitocin usage has been so common in maternity wards in the US for decades, it's even possible some boards have OB/GYNs who have abused Pitocin in the past, who might be sympathetic to peers, or want to protect themselves from past malpractice for their use of the drug. It's not easy to predict the kind of support Pitocin abuse cases might get from state medical boards if they are called on for help. Even though the labeling on Pitocin is clear, even other 'experts' in the field that a court might call upon could be compromised.

The Obstacles to Autism Research

We live in the age of information, where data has come to be the foundation of all logical reasoning. It could easily turn into the golden age of healthcare if data pertaining to millions of individuals can be mined to find the cause and effect of numerous conditions, their treatments and effectiveness of medications and results. Tech companies with the best of intentions have tried making an impact in healthcare, but to no avail. While hospitals are big businesses, they are mostly not billion-dollar corporations capable of putting in a lot of money toward their IT infrastructure. Hospitals with their smaller and tight budgets are torn between maintaining their IT systems and paying for doctors and medical equipment and still turn a profit. Hospital IT systems continue to be locked-in by a select group of vendors with their proprietary systems that restrict the adoption of next generation IT systems. Strict privacy requirements have also meant there has never been any incentive for data sharing for any purpose whatsoever. Billing systems are truly the only IT systems with any decent data in hospitals. They may have disease codes and treatments, but deriving meaningful history of patients from invoices is next to impossible. There is no single personal health record that even doctors can access. Add to this the fact that people change care providers or consult multiple providers and specialists.

The result is it is impossible to decipher meaningful information about individual disease histories and treatments and arrive at correlations that might be beneficial to all and certainly in a case like this. Government initiatives in the US like HITECH, EHR mandate, etc. are steps in the right direction, but they have only scratched the surface when it comes to creating systems capable of sharing data in any meaningful way.

Of course, on the flip side, insurance companies have been trying to get their hands on consolidated patient history information for ages, but for their totally devious purposes of jacking up rates or denying coverage at the slightest hint of disease. This has disincentivized individuals from sharing their information, and as a result stifling any possibility of making data available for research. While individual patients can request their records from various places and put a case together, it is not possible to seek out impacted cases from hospital records from any other angle. This makes discovery in any legal case so much harder as it is almost impossible to gather data to prove a pattern of malpractice by a given doctor.

Autism research has instead depended exclusively on observational studies. Unfortunately, this has been the reason it has taken this long to pinpoint the causes of autism. These are not clinical trials of drugs but passive observations collected over time from various sources. Running a medical study involves a lot of time and resources. They are mostly done by academic researchers and funded by industry with specific goals in mind. With autism, the problem itself is not clearly defined. With the possible vaccine connotation, it's the last thing anybody in the field wants to touch, simply because nobody wants to dive into a field so controversial, where coming up with any conclusions only puts you at odds with the vast majority of other researchers.

Proving medical malpractice and getting into legal complications is something any academic or scientific research program would want to avoid at all costs. Hence even if they touch on the topic, their results always come up as inconclusive and end up pointing to the need for further research. As a result, autism studies have been focused on what could be the most non-controversial cause, which is finding genetic roots for autism. When a deficit in the body is not traceable to viral or bacterial disease, looking for genetic reasons seems to be the default fallback for contemporary science and unfortunately, autism has fallen into that same rut.

With their limited scope and vision, studies trying to pinpoint a single reason for autism are doomed to come up empty-handed – as they have been for the past couple of decades. They are fundamentally flawed because they are studying something with too many different causes and too many consequences.

On the flip side of medical research is the fact that drugs need to fail quickly if they are dangerous. They should not be allowed to cloak themselves under the complexity of the medical system or the complacency of medical personnel or the whims of insurance companies for that matter. When this happens, they continue their carnage, until the effects have impacted way too many people – the way Pitocin has for over 40 years. Current privacy regulations and other restrictions have in this case been a deterrent to the truth emerging earlier.

The Perpetrators

Meaning well does not equate to doing good. Unfortunately, in the case of Pitocin abuse and brain damage, numerous OB/GYNs – in their zeal to avoid the 'dreaded C-section' – have ended up causing autism. Their incompetence and lack of understanding of the dangers of Pitocin induction and inability to direct the treatment as per the labeling on the drug being administered has caused an epidemic. They tend to believe that Pitocin being a naturally occurring hormone is 'safe and the most natural way' to deliver a baby. Their ill-conceived notion that neuroplasticity would help babies recover from any brain damage has resulted in thousands of victims. By their negligence and beliefs contrary to the drug labeling, each of these autism perpetrators has made suffering the only way of life for hundreds of families they have victimized. They tend to be serial Pitocin abusers impacting multiple families the same way, while their malpractice has remained hidden for years, with some of them even rising in power and stature to become part of the powerful medical board committees that wield ultimate power in their fields.

Medical insurance companies that try to push doctors to do fewer C-sections are part of the problem too. Their primary motive is the cost factor that comes with C-sections compared to regular vaginal deliveries. The cost of a typical hospital delivery in the US is almost impossible to guess, with estimates for 2019 ranging from $15,000 to $32,000 for normal deliveries and $30,000 to $52,000 for C-sections (these numbers are the amount hospitals charge insurance companies and most people with insurance end up paying much less). Being one of the most common procedures done in hospitals it's a pity the price would be so high in the first place, but suffice to say C-sections cost approximately 50% more than a normal delivery due to anesthesia, surgery and additional care expenses involved. I see stories on the internet where insurance companies have resorted to shaming obstetricians for doing more C-sections in an effort to curb costs while claiming better outcomes for mothers. Unfortunately for parents of babies impacted by Pitocin abuse, the price of trying to fix the mistake of not doing a C-section runs into millions over the lifetime of the child while the affliction still cannot be fixed. A kid on one-on-one ABA therapy costs approximately $120,000 per year, and that's not counting the pain, suffering and other incidentals that parents and the child have to go through. The long term costs of supporting a kid with autism far exceeds anything insurance companies might have paid for a few additional C-sections.

But the primary culprits in this story are the people who have engaged in manipulation to hide the risks from the messaging. Take for example what happened in 2013 when a large widely reported study from the Duke Center for Autism by Simon Gregory and his colleagues[17.1] for the first time established a correlation between induced labor and an autism diagnosis by analyzing North Carolina birth records with educational records. Their conclusion of the association between augmented labor and autism is perfectly on the money, but in their explanation they seemed to have ascribed plausible cause to a totally flawed hypothesis in the autism research world that oxytocin exposure of the fetal brain could somehow be responsible. Unfortunately, this nonsensical hypothesis seems to have hijacked conventional scientific belief, despite the fact that oxytocin doesn't last in the maternal blood stream for any significant time (that's the reason you have to use it as a continuous drip to induce contractions). Anything that reaches the fetal brain is potentially 1000's of times diluted, and whatever does find its way there is blocked by the blood brain barrier.

It was obviously not possible for the Gregory study to look at the malpractice involved due to deficits in the data they looked at. Nevertheless, there's been an almost immediate torrent of misdirected rebuttals of these results from other studies, with many erroneously faulting the study for pointing fingers at augmented labor instead of sticking to the oxytocin chemical effects thesis. Somehow it doesn't occur to these experts that this could be traumatic brain injury by the force of contractions, which the fetal monitors – designed to pick up hypoxia – do not detect.

What is alarming is the volume of immediate counter studies and rebuttals which seem to have been established solely for the purpose of discrediting the original Duke study. OB/GYN organizations like the ACOG (www.acog.org) seem to be all too eager to embrace these counter arguments that egregiously go overboard defending induced labor.

The case against Pitocin abuse is complex simply because of the numerous other factors that can impact the brain. While fingers have been pointed at induced labor multiple times in the past, nothing has come out of it. However, the time for redemption has arrived. More and more people are now able to trace their kid's brain issues to their birth. My expectation is that this book will cement their knowledge and take away the diversion of vaccines and environmental factors since they are truly a secondary add-on effect arising out of their interaction with a damaged brain with a compromised blood brain barrier.

It is time for impacted parents to lose their dependence on 'studies' and 'experts' and trust what they see with their own eyes and experience. What is needed is a movement by impacted parents to clear the smokescreen once and for all and stop the genetic nonsense being parlayed about autism for years. Outsiders that are not impacted by autism cannot understand it, the way the parents of impacted individuals do. Only impacted parents can bring out the truth that most autism is caused by birth trauma sometimes naturally and at other times by the Pitocin drug, which is living up to its warning label, causing traumatic brain injury at the hands of OB/GYNs lacking the expertise to handle it.

Key Take Away: Pitocin abuse should be easy to establish, but it is a difficult undertaking nevertheless, due to the complex nature of the cases and the backlash to past successes in courts. Heavy-handed tort reform as response to past attempts has resulted in a loss of moral courage to take on these cases among the legal community. The unwillingness of the scientific community to delve into issues of malpractice and their eagerness to embrace genetic reasons to autism that do not exist, has not been very helpful.

CHAPTER 19

Establishing Serial Pitocin Abuse

Considering that it is an uphill task to prove individual cases in court, a faster and easier way to prove malpractice would be to prove multiple instances of Pitocin abuse by a single practitioner. At AutismPI, our approach is to establish class action lawsuits against individual OB/GYN practitioners with patterns of abuse. Our basic premise is that these Pitocin abusers are serial perpetrators of abuse and have caused multiple cases of autism by not respecting the drug's contra-indication limits. Our data collection initiative is an effort to establish this relationship thus providing much-needed proof of serial Pitocin abuse.

The Court Cases and Precedents

In talking to some parents, I sense a certain anxiety around participating in a lawsuit, particularly given all the active message management on the internet. There is also concern around the time something like this might eat up, taking their attention off their kids. Special needs families are already fighting the tough battle of a shortage of time and money. Some turn to God and some of them are beaten down. Some are so crushed they lack the courage to pursue a lawsuit on their own. Then there are some others that are quite the opposite, waiting on their wings for this book and the data collection so they can move ahead with a lawsuit. Understanding the history of past autism lawsuits can help explain the strategy of pursuing a class action lawsuit and why it could succeed where others did not.

Vaccine-Autism Omnibus hearings: Beginning with British gastroenterologist Dr. Andrew Wakefield's 1998 publication of case studies implicating the MMR vaccine with the rise of developmental disorders including autism and inflammatory bowel disease in children, the notion that vaccines were the cause of the astronomical rise of autism since the 1980s started gaining traction. However, several subsequent epidemiological studies debunked his theory, and Dr. Wakefield was dealt with firmly by the British medical establishment, and his publisher (the respected Lancet journal) was forced to retract his statements in 2004. Due to his unflinching advocacy of his findings, after years of deliberation Dr. Wakefield was ultimately stripped of his medical license in 2010. But the popularity of the theory among the general population only increased due to it being corroborated by actual parental observations.

The "Defeat Autism Now" (DAN) movement of Dr. Rimland had accepted the notion that vaccines were the cause of autism, particularly the mercury in the thimerosal adjuvant used as both a preservative and enhancer to improve the effectiveness of vaccines. Mercury poisoning can lead to brain damage and cause symptoms similar to autism and the correlation sounded pretty plausible. This resulted in over 5600 claims being filed against the National Vaccine Injury Compensation Program (NVICP). The vaccine court selected 6 test cases to represent the different circumstances of usage, including vaccines containing thimerosal, MMR and combinations of the two. These consolidated cases constituted the Omnibus Autism Proceedings (OAP) that began in 2007. The court found no evidence of any causative relationship to autism in any of the cases and consequently no compensation was paid in any of the 5600 cases when they were adjudicated in 2009–10, and the future of making a vaccine autism relationship in vaccine court looks bleak.

To make matters worse, the experts called on by the court and the special masters at the Omnibus proceedings argued future research needed to focus on genetics and that research into environmental causes should only be considered as part of gene–environmental interactions. This ended up badly damaging the reputations of anti-vaxxers and their sympathizers among the medical community and the general population while redirecting valuable research dollars into looking for genetic causes that don't exist. As a result, the underlying cause of most autism – medical malpractice – remains "hidden" to this day because it is not even remotely related to genes or the environment.

John Edwards and Electronic Fetal Monitoring: The advent of electronic fetal monitoring in the 1960s opened up the possibility of detecting hypoxia happening during birth. Electronic fetal monitors are the reason induction of labor is even possible. As part of managed labor in a hospital, fetal monitoring has since become standard practice in the US and most other nations. The cardiotocograph uses a monitor placed on the mother's abdominal wall to monitor contractions and fetal heart rate. The fetal heart rate typically drops with every contraction and quickly recovers in the interval between contractions. Fetal distress can be detected when the heart rate doesn't recover quickly enough or stays low for a longer time interval. Yet as straightforward as it sounds, there are complexities that need considered, including a moving baby, change in monitor position, the complexity of signals getting through the abdominal and uterine wall, and so on. A second more reliable but more invasive monitor is the fetal scalp electrode (FSE) which can be used after at least 3cm dilation of the cervix is achieved. The scalp electrode is placed under the fetal skin in the head and among other risks carries the possibility of causing an infection but is credited to be more reliable. But even this apparently is not considered a very reliable predictor of fetal hypoxia.

John Edwards, former US Senator and one-time presidential and vice-presidential candidate, was a birth injury lawyer who excelled in winning medical malpractice claims for his victim clients. His cases were mostly about kids that ended up with cerebral palsy. Some of his wins are subjects of controversy, particularly since the medical community now believes cerebral palsy begins before birth and could be the result of maternal infections and other causes. John Edwards is said to have successfully argued his cases based on the electronic fetal monitor readings.

Furthermore, he went one step further and not only sued the doctors, whose malpractice insurance settled the claims, but also the hospitals for letting this happen in their facilities, holding them liable for millions in additional compensation. Wikipedia estimates John Edwards made 38 million dollars for himself while winning 208 million for his various clients in the 1980's.

However, the John Edwards suits may have brought about heavy-handed tort reform by way of many states limiting jury awards and putting time limits of less than 18 years for birth injury lawsuits. These ill-advised measures have further complicated the possibility of winning malpractice suits for birth injury. People have been accused of pointing to a squiggly in electronic fetal monitor graph as cause for brain damage in their baby and claiming compensation. Numerous subsequent cases against prolonged induced labor have since been lost in courts. Unfortunately, this has emboldened a new generation of OB/GYNs who in their excessive zeal to please insurance companies have brought on Pitocin abuse, raising autism levels to unprecedented highs.

The Autism Parents Initiative: Autismpi.org

Proving medical malpractice as the true cause of autism needs to be based on solid foundations and irrefutable evidence. Due to the enormous diversity of the conditions and mechanisms causing brain damage at birth, standardized criteria need to be established on what constitutes Pitocin abuse. While the contra-indication criterion is clear, the outcomes are varied. Data needs to be gathered at a large scale, possibly from every impacted parent on the various outcomes that have been seen. Data is the new gold in the digital world, and mining for this gold will establish clear patterns of Pitocin abuse and outcomes.

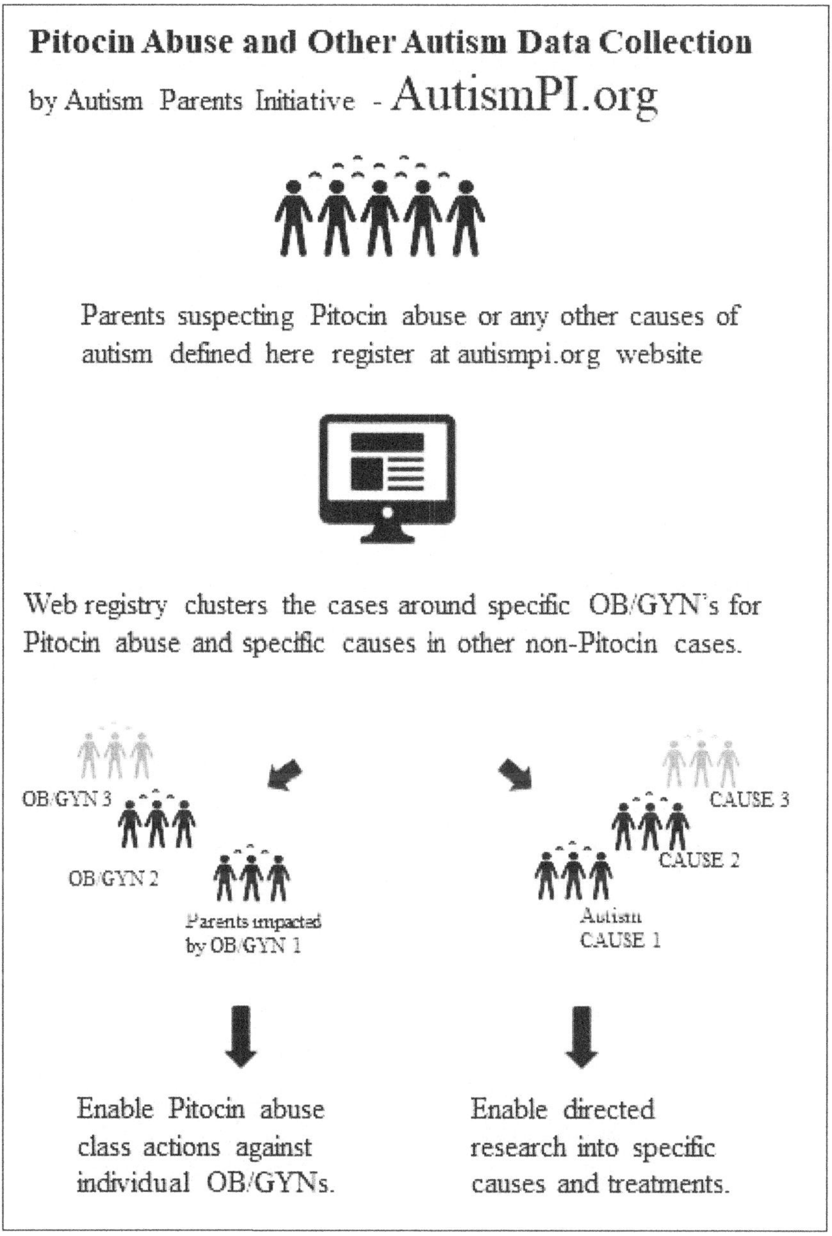

The Autism Parents Initiative or AutismPI (pronounced autism π – autismpi.org) is an effort by a group of autism impacted parents (including me) that have created an organization for help with tracking down causes of autism, and in cases of Pitocin abuse help with formation of legal class action suits against Pitocin abuse by individual OB/GYNs. We have a website to serve as an online Pitocin abuse injury registry to register your case details. This is a data collection and information dissemination effort to help successful legal initiatives against Pitocin injury. We intend to collect this information and report back to you when we find clusters of autism by Pitocin abuse against your OB/GYNs, which gives you the grounds for a successful class action lawsuit. Our goal is to help every Autism parent that has been similarly impacted by Pitocin abuse in induced labor demand and receive compensation for the injustice meted out to them and their children.

Given the history of trying to prove autism in court, a legal case against a medical practitioner for Pitocin abuse works best if multiple instances of abuse per OB/GYN are identified. Hence, as the first step of our strategy, we seek to identify serial Pitocin abuser gynecologists that have destroyed our lives and our children's. In order to keep out vested interests from sullying the data, we collect a nominal fee so we can cross-reference card payment information to keep the accuracy of the information and put in some more time and effort in what is otherwise an open website that anybody can register in with just an email. We hope to run this data collection campaign for up to 2 years in the US. The site is designed to eliminate the last barrier between parents and that elusive proof required to prove a case in a court of law. Autismpi.org is a quest to prove autism by Pitocin abuse by the numbers, backing up the conclusions arrived at in this book using plain logical arguments.

We have also begun putting together a network of interested lawyers by region in the US. that have interest and expertise in pursuing these cases. It would be our interest to see these cases succeed and it would be an additional service AutismPI could offer in future should parents opt to take advantage of such a service. It would be based on a lower, pre-negotiated, group discounted contingency fee given the high fees in the industry. Lawyers can typically charge 33 – 40% of the recovery in cases fought on contingency. Most of the legal costs are due to discovery requirements and the AutismPI platform would be able to coordinate some of that discovery through the data collection efforts, thereby reducing overall costs and efforts and pass on those discounts to parents. Lawyers networked through the AutismPI network would have access and support for their efforts and would charge much less as a result of the reduced efforts and better guarantee of success.

Most autism parents would rather ask for their kids to be fixed than seek monetary compensation. But unfortunately for us, the only thing the court system can offer is compensation. There are not any fixes on the horizon that are guaranteed to that work and even if something does work out eventually, there's no returning the years lost in the struggle. Focusing on this legal initiative is important to secure the financial future of these kids and to fix the problems that have been caused by this tragedy. It's essential to put these doctors out of business rather than allow them to continue to practice.

Collection of new data by conducting new surveys from the parent-side can also lead to more focused research and the creation of better treatments for brain damage at birth from hypoxia, ischemia and other natural causes and even assess other impacts of induced labor. For example, do induced mothers have a higher risk of uterine cancer? There is no way to find out unless there is a centralized effort to track the data.

A lack of data is the reason why this medical malpractice has remained hidden, but together we can bring down this last barrier to ending autism in this world. I appeal to impacted parents to register today so together we can bring the curse of Autism to an end and seek compensation for the medical malpractice that has been hiding in plain sight for almost 4 decades.

Key Take Away: Pitocin contra-indication and doctors exceeding labelled limits ought to be a straightforward and clear-cut legal case. Yet, years of tort reform, have made things a lot more complicated. Establishing a pattern of abuse against individual obstetricians makes these cases a lot easier. AutismPI.org is a parents' organization dedicated to collecting data on autism causes and of all pathways with strong advocacy for brain damage and regeneration research. AutismPI.org is also running a data collection initiative to catalog causes of autism. This data collection is also aimed at clustering malpractice data on obstetricians and hospitals, thus making the cases a lot easier to pursue as class actions against individual practitioners and hospitals. If you are a parent of a child with autism, it is very important to register your data with AutismPI.org.

CONCLUSION

The Path Forward

Autism has been a problem that has confounded the scientific community for decades now. By now you probably understand why. When you realize a problem of such enormity is caused by medical malpractice and drug injury as the predominant cause, it can send you raging with anger for a long time. Hammering a baby out of the womb headfirst against a non-effacing cervix using a drug beyond the point of labeled drug contra-indication is malpractice of epic proportions. At least 25% of all births in the US are known to be induced. There are over 3 million individuals suffering from autism in the US alone, with a high percentage of this being traumatic brain injury at birth caused by badly mismanaged induced labor. This is a crime against unborn children, virtually ruining their lives for good.

I have shared our story of how a clueless OB/GYN turned a perfectly healthy 'low risk' pregnancy into a case of Autism not once but twice in our life. I have shared the evidence of damage and the clear evidence of malpractice where the OB/GYN went against established norms at the hospital causing MRI visible brain damage. And I have met numerous other parents with similar stories – in fact it's a story that is repeated all too often in maternity wards. It's a story only impacted parents would know from start to finish, from birth to the autism diagnosis. The best judges of this situation are the impacted parents, but many of them in their desperation to help their children have become overly dependent on the scientific community to give them the answers. They are forgetting that as the people spending the most time with these kids, they have front row seats to years of observation that science is not privy to. The scientific community has been caught in conflicting professional and economic interests and the results have been confusing, leading to a deliberate lack of clarity in conclusions of studies for fear of repercussions to any statements they make on this emotionally charged topic. Privacy laws have worked out in favor of some special interests to further stifle the truth for years.

I have shown how the traumatic brain damage has remained hidden in most cases due to damage at the first cervical dilation stage of labor being literally 'erased' by the pushing stage of labor as seen in the MRI evidence of my own boys. We have seen how the damaged blood brain barrier leads to gut dysbiosis and mitochondrial dysfunction in many cases of autism and how the same symptoms are shared even by adult victims of TBI. I have also reasoned how live virus vaccines and for that matter many other viruses might bring about these effects in brain damaged kids by way of viral latencies. We have also looked at the contra-indications for Pitocin in its labeling and I have shared my story of exactly how it gets to be abused. All this solid evidence is known to many people but has probably not been said the way I have. Sometimes it takes someone to say things the way they are, before change can happen.

The time has come to end autism for good and the best way I can see to bring this about is for the battle to be taken up in the US legal system (and perhaps serve as the model for other parts of the world). It can only be done by identifying these autism producing gynecologists who are either blissfully unaware of the explicit warning labels on the Pitocin drug or totally lack understanding of the consequences of its contra-indication conditions. I believe every impacted parent reading this book who's been through the induced labor process has a similar story to tell. Our stories are real, but in the face of schemes to thwart us from finding the truth we will need to band together to bring out the truth.

It is no longer necessary for autism parents to accept autism as a fact of life that happened to them. With this book, I have provided the real reasons for this scourge. Doing away with the practice of using induced labor after amnion rupture and putting C-section back as the main and only alternative will eradicate a signification portion of autism. That will only happen if parents rally around and identify the gynecologists responsible for serial malpractice and bring them to justice while seeking compensation for the irreparable damage caused to the lives of these kids. Clustering autism data by OB/GYN and hospital can allow us to isolate specific OB/GYN's that are routine Pitocin abusers responsible for large numbers of autism cases. The data collection effort by AutismPI.org (pronounced autism pie/π) is designed to bring out the evidence of malpractice against individual practitioners and their hospitals.

There are bound to be people trying to put down our voices and discredit our analysis, perhaps even some with great credentials. What does a software engineer know that we in the research world do not understand? But this is a topic where researchers have been looking in the wrong places, at the wrong types of data and coming up empty-handed for ages. Parents being the closest observers of their kids are the best judges of their individual situations.

As you may have realized, this book is not just about the ideas that have been put forward. There is going to be action to back it up. To help eradicate the scourge of autism and to seek compensation for impacted parents' woes, it is important for impacted parents to band together and pursue legal recourse.

I invite you join the autism parent initiative today. AutismPI is a no-obligation data collection effort for all autism in the U.S. aimed at cataloging the various birth causes of autism as seen by parents and in cases of Pitocin abuse a separate registry is intended to prove medical malpractice at birth by Pitocin abuse. Register your Pitocin abuse case at our parents founded organization at https://autismpi.org and participate in bringing out the overwhelming evidence against Pitocin abuse in induced labor and help identify your OB/GYN for the autism producer they really are. The evidence could make the difference in a class action lawsuit against your individual obstetrician.

If you have any doubts at all, a look at your children's eyes will tell you they are so worth it. They were deprived of their basic human abilities by these practitioners unworthy of handling the responsibility they were trained to execute. It's upon us impacted parents to stand up to the injustice and save the future from ongoing malpractice and the scourge of autism.

Let's
mark a new
beginning
of the end
to the
malpractice malady
we call
"Autism"

References

Too many research papers out there seem to start with the faulty premise that all autism is genetic. Such sweeping generalization makes many of these 'autism' papers irrelevant. But good quality research is invaluable and I am grateful for some very good information and images made available by many educational institutions, nonprofits and some select papers available from the National Institute of Health website. Some of these references used in individual chapters are listed below.

1.1) Benvenuto A, Manzi B, Alessandrelli R, Galasso C, Curatolo P. Recent advances in the pathogenesis of syndromic autisms. *Int J Pediatr.* 2009;2009:198736. doi:10.1155/2009/198736

1.2) Konkel L. The brain before birth: using fMRI to explore the secrets of fetal neurodevelopment. Environ Health Perspect. 2018;126:112001. https://ehp.niehs.nih.gov/doi/10.1289/EHP2268

1.3) Stiles J, Jernigan TL. The basics of brain development. Neuropsychol Rev. 2010;20(4):327-348. Doi:10.1007/s11065-010-9148-4

1.4) Paus T, Zijdenbos A, Worsley K, et al. Structural maturation of neural pathways in children and adolescents: in vivo study. Science. 1999;283(5409):1908-1911. Doi:10.1126/science.283.5409.1908

3.1) Saunders NR, Liddelow SA, Dziegielewska KM. Barrier mechanisms in the developing brain. Front Pharmacol. 2012;3:46. Published 2012 Mar 29. doi:10.3389/fphar.2012.00046.

3.2) Sudo N, Chida Y, Aiba Y, et al. Postnatal microbial colonization programs the hypothalamic-pituitary-adrenal system for stress response in mice. *J Physiol.* 2004;558(Pt 1):263-275. doi:10.1113/jphysiol.2004.063388

4.1) Changes in Practice among Physicians with Malpractice Claims, Publication: The New England Journal of Medicine, Publisher: Massachusetts Medical Society, Date: Mar 28, 2019

6.1) Wolf JH. Risk and Reputation: Obstetricians, Cesareans, and Consent. J Hist Med Allied Sci. 2018;73(1):7-28. Doi:10.1093/jhmas/jrx053 – https://www.ncbi.nlm.nih.gov/pmc/articles/PMC5892390/

9.1) Wilcox CL, Johnson EM. Characterization of nerve growth factor-dependent herpes simplex virus latency in neurons in vitro. J Virol. 1988;62:393–399.

9.2) Siddiqui MF, Elwell C, Johnson MH. Mitochondrial Dysfunction in Autism Spectrum Disorders. Autism Open Access. 2016;6(5):1000190. Doi:10.4172/2165-7890.1000190

9.3) Emad M. et al. (2017). The soft mechanical signature of glial scars in the central nervous system. *Nature Communications.*

9.4) Angoa-Pérez, M., Zagorac, B., Anneken, J.H. *et al.* Repetitive, mild traumatic brain injury results in a progressive white matter pathology, cognitive deterioration, and a transient gut microbiota dysbiosis. *Sci Rep* **10**, 8949 (2020). https://doi.org/10.1038/s41598-020-65972-4

9.5) Zhu CS, Grandhi R, Patterson TT, Nicholson SE. A Review of Traumatic Brain Injury and the Gut Microbiome: Insights into Novel Mechanisms of Secondary Brain Injury and Promising Targets for Neuroprotection. *Brain Sci.* 2018;8(6):113. Published 2018 Jun 19. doi:10.3390/brainsci8060113

10.1) Peesay M. Nuchal cord and its implications. Matern Health Neonatol Perinatol. 2017;3:28. Published 2017 Dec 6. Doi:10.1186/s40748-017-0068-7

11.1) Cornette L. Therapeutic hypothermia in neonatal asphyxia. *Facts*

Views Vis Obgyn. 2012;4(2):133-139.

13.1) Jolles S, Sewell WA, Misbah SA. Clinical uses of intravenous immunoglobulin. Clin Exp Immunol. 2005;142(1):1-11. doi:10.1111/j.1365-2249.2005.02834.x

13.2) Lee RWY, Corley MJ, Pang A, Arakaki G, Abbott L, Nishimoto M, Miyamoto R, Lee E, Yamamoto S, Maunakea AK, Lum-Jones A, Wong M. A modified ketogenic gluten-free diet with MCT improves behavior in children with autism spectrum disorder. Physiol Behav. 2018 May 1;188:205-211. doi: 10.1016/j.physbeh.2018.02.006. Epub 2018 Feb 5. PMID: 29421589; PMCID: PMC5863039.

14.1) Boese, A.C., Le, Q.E., Pham, D. et al. Neural stem cell therapy for subacute and chronic ischemic stroke. Stem Cell Res Ther 9, 154 (2018). https://doi.org/10.1186/s13287-018-0913-2

14.2) Shoichet MS, Tate CC, Baumann MD, et al. Strategies for Regeneration and Repair in the Injured Central Nervous System. In: Reichert WM, editor. Indwelling Neural Implants: Strategies for Contending with the In Vivo Environment. Boca Raton (FL): CRC Press/Taylor & Francis; 2008. Chapter 8. Available from: https://www.ncbi.nlm.nih.gov/books/NBK3941/

14.3) Li H, Chen G. In Vivo Reprogramming for CNS Repair: Regenerating Neurons from Endogenous Glial Cells. Neuron. 2016;91(4):728-738. Doi:10.1016/j.neuron.2016.08.004

14.4) Gao, X., Wang, X., Xiong, W. et al. In vivo reprogramming reactive glia into iPSCs to produce new neurons in the cortex following traumatic brain injury. Sci Rep 6, 22490 (2016). https://doi.org/10.1038/srep22490

14.5) Sun YJ, Zhang ZY, Fan B, Li GY. Neuroprotection by Therapeutic Hypothermia. Front Neurosci. 2019;13:586. Published 2019 Jun 11. 9.

14.6) Hasan A, Deeb G, Rahal R, Atwi K, Mondello S, Marei HE, et al. Mesenchymal stem cells in the treatment of traumatic brain injury. Front Neurol. (2017) 8:28. doi: 10.3389/fneur.2017.00028

14.7) Zibara K, Ballout N, Mondello S, Karnib N, Ramadan N, Omais S, et al. Combination of drug and stem cells neurotherapy: potential interventions in neurotrauma and traumatic brain injury. Neuropharmacology. (2019) 145:177–98. doi: 10.1016/j.neuropharm.2018.09.032doi:10.3389/fnins.2019.00586

14.8) Maas AIR, Menon DK, Adelson PD, Andelic N, Bell MJ, Belli A, et al. Traumatic brain injury: integrated approaches to improve prevention, clinical care, and research. Lancet Neurol. (2017) 16:987–1048. doi: 10.1016/S1474-4422(17)30371-X

15.1) Macones GA, Hankins GD, Spong CY, Hauth J, Moore T. The 2008 National Institute of Child Health and Human Development workshop report on electronic fetal monitoring: update on definitions, interpretation, and research guidelines. Obstet Gynecol. 2008 Sep;112(3):661-6. doi: 10.1097/AOG.0b013e3181841395. PMID: 18757666. https://pubmed.ncbi.nlm.nih.gov/18757666/

15.2) Lothian, Judith A.. "Saying "No" to Induction." The Journal of Perinatal Education vol. 15,2 (2006): 43–45. doi:10.1624/105812406X107816 https://www.ncbi.nlm.nih.gov/pmc/articles/PMC1595289/

15.3) Budden A, Chen LJ, Henry A. High-dose versus low-dose oxytocin infusion regimens for induction of labour at term. Cochrane Database Syst Rev. 2014 Oct 9;(10):CD009701. doi: 10.1002/14651858.CD009701.pub2. PMID: 25300173. https://pubmed.ncbi.nlm.nih.gov/25300173/

15.4) Bishop, Edward H. (August 1964). "Pelvic Scoring for Elective Induction". Obstetrics & Gynecology. 24 (2): 266–268. PMID 14199536.

15.5) Tenore J (2003). "Methods for cervical ripening and induction of labor". Am Fam Physician. 67 (10): 2123–8. PMID 12776961.

16.1) Nocon JJ, Coolman DA. Perinatal malpractice. Risks and prevention. J Reprod Med. 1987;32(2):83-90.

16.2) Karl SJ Oláh FRCOG FRCS MRCP(I),a,* Philip J Steer BSc MD FRCOGb The use and abuse of oxytocin

16.4) Young RC, Barendse P. Linking myometrial physiology to intrauterine pressure; how tissue-level contractions create uterine contractions of labor. PloS Comput Biol. 2014;10(10):e1003850. Published 2014 Oct 16. Doi:10.1371/journal.pcbi.1003850

16.5) Sanchez-Ramos, L, Kaunitz, A, Glob. Libr. Women's med., (ISSN: 1756-2228) 2009; DOI 10.3843/GLOWM.10130 (The global library of women's medicine)

16.7) Cummins G, Kremer J, Bernassau A, et al. Sensors for Fetal Hypoxia and Metabolic Acidosis: A Review. Sensors (Basel). 2018;18(8):2648. Published 2018 Aug 13. Doi:10.3390/s18082648

16.8) https://www.acog.org/Clinical-Guidance-and-Publications/Committee-Opinions/Committee-on-Obstetric-Practice/Labor-Induction-or-Augmentation-and-Autism

16.9) Sequeira S, Ahmed M. Meditation as a potential therapy for autism: a review. Autism Res Treat. 2012;2012:835847. Doi:10.1155/2012/835847

16.10) Siniscalco D, Kannan S, Semprún-Hernández N, Eshraghi AA, Brigida AL, Antonucci N. Stem cell therapy in autism: recent insights. Stem Cells Cloning. 2018;11:55-67. Published 2018 Oct 23. Doi:10.2147/SCCAA.S155410

16.11) Boese, A.C., Le, Q.E., Pham, D. et al. Neural stem cell therapy for subacute and chronic ischemic stroke. Stem Cell Res Ther 9, 154 (2018). https://doi.org/10.1186/s13287-018-0913-2

16.12) Shoichet MS, Tate CC, Baumann MD, et al. Strategies for Regeneration and Repair in the Injured Central Nervous System. In: Reichert WM, editor. Indwelling Neural Implants: Strategies for Contending with the In Vivo Environment. Boca Raton (FL): CRC Press/Taylor & Francis; 2008. Chapter 8. Available from: https://www.ncbi.nlm.nih.gov/books/NBK3941/

16.13) Li H, Chen G. In Vivo Reprogramming for CNS Repair: Regenerating Neurons from Endogenous Glial Cells. Neuron. 2016;91(4):728-738. Doi:10.1016/j.neuron.2016.08.004

16.14) Gao, X., Wang, X., Xiong, W. et al. In vivo reprogramming reactive glia into iPSCs to produce new neurons in the cortex following traumatic brain injury. Sci Rep 6, 22490 (2016). https://doi.org/10.1038/srep22490

16.15) Jolles S, Sewell WA, Misbah SA. Clinical uses of intravenous immunoglobulin. Clin Exp Immunol. 2005;142(1):1-11. doi:10.1111/j.1365-2249.2005.02834.x

17.1) Gregory SG, Anthopolos R, Osgood CE, Grotegut CA, Miranda ML. Association of autism with induced or augmented childbirth in North Carolina Birth Record (1990-1998) and Education Research (1997-2007) databases. JAMA Pediatr. 2013;167(10):959-966. doi:10.1001/jamapediatrics.2013.2904

17.2) Kurth L, Haussmann R. Perinatal Pitocin as an Early ADHD Biomarker: Neurodevelopmental Risk? Journal of Attention Disorders. 2011;15(5):423-431. doi:10.1177/1087054710397800

17.3) Smallwood M, Sareen A, Baker E, Hannusch R, Kwessi E, Williams T. Increased Risk of Autism Development in Children Whose Mothers Experienced Birth Complications or Received Labor and Delivery Drugs. ASN Neuro. 2016;8(4):1759091416659742. Published 2016 Aug 9. Doi:10.1177/1759091416659742

18.1) Decoteau CL, Underman K. Adjudicating non-knowledge in the Omnibus Autism Proceedings. Soc Stud Sci. 2015;45(4):471-500. Doi:10.1177/0306312715600278

18.2) Decoteau, C. L., & Daniel, M. (2020). Scientific Hegemony and the Field of Autism. American Sociological Review, 85(3), 451-476. doi:10.1177/0003122420922531

More References

Image Sources and Credits.
Images have been predominantly sourced from Wikimedia, Government and Educational websites and public domain. Some images are created by the author

1.1: Brain development with age - Smithsonian Dept of Human Origin, by Karen Carr Studios.

1.2: Brain functional development - https://developingchild.harvard.edu/resources/inbrief-science-of-ecd

1.3: Structural parts of the brain - Image modified by author from purchased material.

1.4: Cerebral functional areas - Image created by author.

2.1: Neuron cell - Wikipedia, "Anatomy and Physiology" by the US National Cancer Institute's Surveillance, Epidemiology and End Results (SEER)

2.2: Neuron 3D - Wikimedia Commons: public domain image by Ladyofhats

2.3: Neuroglial Cells - Wikimedia Commons: Artwork by Holly Fischer labelled by author.

2.4: Grey matter -Wikimedia Commons – from public domain.

2.5: Brain cross section – from public domain - re-labelled by author.

3.1: Blood Brain Barrier – By Author

3.2: Immune system – By Author

4.1: Barriers to birth – By author

4.2: Labor feedback loop - (Image source: OpenStax - https://cnx.org/contents/FPtK1zmh@8.25:fEI3C8Ot@10/Preface, CC BY 4.0, https://commons.wikimedia.org/w/index.php?curid=30131134)

4.3: Three stages of labor - http://mstcparamedic.pbworks.com/w/page/21902870/Stages of Labor

5.1a: Brain sectional planes - Blausen.com staff (2014). "Medical gallery of Blausen Medical 2014". *WikiJournal of Medicine* 1 (2): 10. doi:10.15347/wjm/2014.010. ISSN 2002-4436;

5.1b: Brain sectional planes - case.edu/med/neurology/NR/MRI Basics.htm

7.1: Image by author

9.1: From actual MMR label from internet sources.

9.2: Figure by author.

10.1: Figure by author

10.2: Sourced from CDC website at https://www.cdc.gov/ncbddd/birthdefects/microcephaly.html

10.3: Cord complications - Image by author

10.4: Placenta abruption - Image by author.

10.5: Pitocin abuse cycle – Figure by author.

11.1a: Brain Gyrification 1 – Cowan WM. The development of the brain. Sci Am. 1979 Sep;241(3):113-33. PMID: 493917.

11.1b: Brain Gyrification 2 – Image by author.

11.2: Brain ability to learn - Center on the Developing child. Harvard University. Data source Levitt (2009).

13.1: Dan protocol – Internet sources, from page 67 of the book "*Autism: Effective Biomedical Treatments*"

16.1: Uterus – sourced and labelled by author from https://upload.wikimedia.org/wikipedia/commons/9/96/Blausen_0399_FemaleReproSystem_01.png

16.2: Uterus - public domain image

16.3: Placenta – public domain image with minor modifications by author

16.4: Cardiotocograph – By Martin Hawlish from German Wikimedia Commons
16.5: Amniotic fluid and Induction. - Image by author.
17.1: ACOG website
17.2: Google search result for 'pitocin autism' in 2017
17.3: Pitocin Autism search results on Bing in 2020

Useful Web Links

w1.1	American Pregnancy Association	https://americanpregnancy.org/labor-and-birth/
w1.2	Labor – John Hopkins Medicine	https://developingchild.harvard.edu/resources/inbrief-science-of-ecd
w2.1	Brain Architecture	https://developingchild.harvard.edu/science/key-concepts/brain-architecture/
w2.2	Early Childhood Development	https://developingchild.harvard.edu/resources/inbrief-science-of-ecd
w2.3	Queensland Brain Institute	https://qbi.uq.edu.au/brain/brain-anatomy
w2.4	Transmitting fibers in the brain	https://aiimpacts.org/transmitting-fibers-in-the-brain-total-length-and-distribution-of-lengths/
w3.1	State of World Mothers report	https://www.savethechildren.org/content/dam/usa/reports/advocacy/sowm/sowm-2015.pdf
w5.1	Disease Eradication	https://www.historyofvaccines.org/content/articles/disease-eradication
w5.2	ProCon.org History of vaccines	https://vaccines.procon.org/history-of-vaccines/
w5.3	History of Anti-vaccination Movements	https://www.historyofvaccines.org/content/articles/history-anti-vaccination-movements

w5.4	Timeline-History of Vaccines	https://www.historyofvaccines.org/timeline/all
w5.5	Nerve Growth Factor	https://en.wikipedia.org/wiki/Nerve_growth_factor
w6.1	Infectious Disease in Pregnancy	https://www.merckmanuals.com/professional/gynecology-and-obstetrics/pregnancy-complicated-by-disease/infectious-disease-in-pregnancy
w6.2	The Truth About CPD	https://www.bellybelly.com.au/birth/small-pelvis-big-baby-cpd/
w6.3	Diabetes During Pregnancy	https://www.stanfordchildrens.org/en/topic/default?id=diabetes-and-pregnancy-90-P02444
w6.4	Nuchal Cord:	https://www.thebump.com/a/nuchal-cord
w6.5	Cerebral Palsy	https://www.cerebralpalsy.org/about-cerebral-palsy/cause
w6.6	Hypoxic-Ischemic Encephalopathy	https://www.cerebralpalsy.org/about-cerebral-palsy/cause/hypoxic-ischemic-encephalopathy
w6.7	Dyslexia – Symptoms and causes	https://www.mayoclinic.org/diseases-conditions/dyslexia/symptoms-causes/syc-20353552
w6.8	Birth Defects	https://www.cdc.gov/ncbddd/birthdefects
w7.1	About Oxytocin	https://psychcentral.com/lib/about-oxytocin/
w7.2	What is Pitocin?	https://www.bundoo.com/articles/what-is-pitocin/
w7.3	The Truth About PITOCIN	https://pathwaystofamilywellness.org/Pregnancy-Birth/the-truth-about-pitocin.html
w7.4	5 Things Oxytocin does that Pitocin doesn't	https://www.bellybelly.com.au/birth/5-things-oxytocin-does-that-pitocin-doesnt/
w7.5	Induced Labor Malpractice -	https://www.kennerlyloutey.com/birth-injuries/induced-labor-pitocin/

Index

D

E

F

G

H

P

R

S

T

U

V

W

Z

Autism

Parents Initiative

The journey to the beginning of the end of autism begins with the data collection at www.autismpi.org

Register autism birth data there now to enable research and legal discovery.

Made in United States
North Haven, CT
20 October 2026

20740144R00134